Martin Spiecker

Two-level system hyperpolarization with a quantum Szilard engine

Experimental Condensed Matter Physics
Band 34

Eine Übersicht aller bisher in dieser Schriftenreihe erschienenen Bände finden Sie am Ende des Buchs.

Two-level system hyperpolarization with a quantum Szilard engine

by
Martin Spiecker

KIT Scientific Publishing

Karlsruher Institut für Technologie
Physikalisches Institut

Two-level system hyperpolarization with a quantum Szilard engine

Zur Erlangung des akademischen Grades eines Doktors der Naturwissenschaften von der KIT-Fakultät für Physik des Karlsruher Instituts für Technologie (KIT) genehmigte Dissertation

von Martin Spiecker, M.Sc.

Tag der mündlichen Prüfung: 26. April 2024
Referent: Prof. Ioan Pop
1. Korreferent: Prof. Alexander Shnirman
2. Korreferent: Prof. Stefan Filipp

Impressum

KIT Scientific Publishing

Karlsruher Institut für Technologie (KIT)
KIT Scientific Publishing
Straße am Forum 2
D-76131 Karlsruhe

www.bibliothek.kit.edu/ksp.php | E-Mail: info@ksp.kit.edu | Shop: www.ksp.kit.edu

Print on Demand 2025 – Gedruckt auf FSC-zertifiziertem Papier

ISSN 2191-9925
ISBN 978-3-7315-1415-2
DOI 10.5445/KSP/1000179014

Contents

Abstract

In the last decades, superconducting circuits have made important contributions to the study of quantum mechanical phenomena. Their performance approaches the threshold allowing for fault tolerant quantum computation. However, the innate complexity of solid-state physics exposes superconducting quantum circuits to interactions with uncontrolled degrees of freedom degrading their coherence. Although tremendous progress has been made to improve the coherence of superconducting circuits, they still have to cope with various loss and decoherence mechanisms, and with further improvements, it becomes increasingly challenging to track down individual decoherence mechanisms.

By implementing a quantum Szilard engine with an active feedback control loop, we show that a superconducting granular aluminum fluxonium qubit is coupled weakly to a two-level system (TLS) environment of unknown physical origin, with a relatively long intrinsic energy relaxation time exceeding 50 ms. As part of the hyperpolarization with the quantum Szilard engine, the TLSs can be cooled down, resulting in a four times lower qubit population, or they can be heated up to manifest themselves as a negative-temperature environment. We show that the TLSs and the qubit are each other's dominant loss mechanism and that the qubit relaxation is independent of the TLS populations. Since the TLSs are much longer lived than the qubit, non-exponential relaxation and non-Poissonian quantum jumps can be observed. The incoherent relaxation dynamics of the system is described by the Solomon equations, for which a rigorous derivation is presented starting from a general Lindblad equation for the qubit and an arbitrary number of TLSs. In the limit of large numbers of TLSs, the relaxation is likely to follow a power law, which is deduced from the Solomon equations and confirmed experimentally. Moreover, the measured non-Poissonian quantum jump statistics can be reproduced by a diffusive stochastic Schrödinger equation. With increasing number of TLSs, entanglement and measurement back action can be ignored, and the quantum jump statistics can also be reproduced by the Solomon equations. The transition from a stochastic Schrödinger equation model to the Solomon equations hints at a quantum-to-classical transition.

Part I

Main results

1 Introduction

Understanding the foundations of the world has always been a driving force for mankind. With the advent of the second quantum revolution, heralded by the control of individual quantum states, completely new possibilities are opening up, and chances are high that we can wrest one or two secrets from quantum mechanics. On the technological side, progress is expected from quantum sensing, quantum simulations and quantum computation. The dream of universal quantum computation, which is attracting a lot of attention, will probably still be a while in coming. Nevertheless, qubit decoherence times, as well as readout and gate fidelities are approaching the threshold for quantum error correction. In fact, it has recently been demonstrated on a superconducting quantum hardware that quantum error correction can prolong the lifetime of the computational subspace of the device beyond the lifetimes of all the elements it is composed of [1].

Beyond quantum computing, superconducting quantum hardware, with its remarkable design flexibility, is perfectly suited for fundamental research studies. In recent years, numerous experiments have been presented, spanning from the development of quantum hybrid systems [2, 3], to exploring quantum many-body effects and thermodynamics [4–7], to the study and utilization of quantum measurement processes [8–10]. Superconducting hardware will continue to develop into an exciting research field, regardless of which platform ultimately proves most suitable for quantum computation.

The basic building block in the quantum world is the two-level system. In the context of quantum information processing, it is referred to as the quantum bit or qubit. In contrast to its classical counter part, the classical bit, which can either take on the value 0 or 1, the qubit can be in a complex superposition of its eigenstate $|\psi\rangle = \alpha\,|0\rangle + \beta\,|1\rangle$ with the normalization $|\alpha|^2 + |\beta|^2 = 1$. This superposition, however, does not yet make the qubit special. A superposition of two modes with different amplitudes and phases can also be realized in the classical world [11]. The first main peculiarity of quantum mechanics stems from the fact that the superposition is not confined to a single qubit but extends over all elements in the system and even to the environment of the system. The wave function of N two-level systems or qubits takes on the form

$$|\psi\rangle = c_0\,|11\ldots1\rangle + c_1\,|01\ldots1\rangle + \cdots + c_{2^N-1}\,|00\ldots0\rangle\,, \tag{1.1}$$

with corresponding normalization. This exponential scaling of the state space is what people wish to harness in quantum information.

In this work, the interaction with more than 50 two-level systems (TLSs) is observed. While the corresponding state space can in principle still be handled by modern classical supercomputers, the number of states already exceeds the estimated number of atoms in the visible universe for 300 TLSs. It goes without saying that such a system will never be tackled numerically in its full complexity. On the contrary, it is also clear that with the control over N qubits, it is not possible to address the full Hilbert space in a finite amount of time [12], which is why only a handful of algorithms providing a quantum advantage are known to date [13].

In view of this immense Hilbert space, one may wonder to what extent the rich quantum mechanical dynamics can be observed in experiments, and how our classical reality emerges, especially when contemplating the deterministic evolution of the wave function governed by the Schrödinger equation. A potential explanation could lie in the second main peculiarity of the quantum world, the measurement process. While the concept of quantum non-demolishing (QND) readout led to the development of high fidelity quantum measurements, it is still unclear what exactly happens during the collapse of the wave function, whether the dynamics is still deterministic and governed by a much larger Hilbert space including the measurement apparatus and the outside world or if the collapse is an intrinsic process in addition to the deterministic evolution [14]. Nonetheless, the measured object is in a product state with the reset of the world immediately after the measurement process when realized with high fidelity, before it dissolves again into its surrounding.

In conjunction with a fast feedback loop, the QND measurement process is the key ingredient for implementing the quantum Szilard engine in this work, thereby enabling the hyperpolarization of the heretofore unknown TLS environment interacting with the qubit. The precise energy flow during the hyperpolarization, whether coming from the measurement apparatus itself or from additional environments that are in contact with the qubit [7], eludes current experimental access. Nonetheless, by reproducing the measured non-Poissonian quantum jump statistics with a diffusive stochastic Schrödinger equation, it will be shown that the qubit and the TLSs become entangled. However, the quantum jump statistics approaches the classical expectation with increasing system size and qubit relaxation, which may be viewed as a quantum-to-classical transition. Moreover, when it comes to the average incoherent relaxation dynamics of the qubit and the TLSs, the complexity can be reduced to the N-dimensional Solomon equations, which are adapted in this work from the nuclear magnetic resonance (NMR) community to the field of quantum information.

The main focus of this thesis is the implementation of granular aluminum fluxonium qubits and the study of their performance. A fluxonium consists of a Josephson junction that is shunted by a large inductance, the so-called superinductance [15]. In this work, the superinductance is realized with the high kinetic inductance of granular aluminum. In this way, the coherence properties of granular aluminum [16–18] can be studied conveniently in the single photon regime. It will be shown that using granular aluminum in superconducting quantum devices allows the implementation of compact designs, from superconducting qubits [19], over readout resonators to microwave filters. In addition, the magnetic field resilience of granular aluminum [20] will be valuable for realizing hybrid architectures consisting of fluxoniums in conjunction with spins.

The thesis is divided into two parts. The first part presents the main results, while the second part provides further detailed information on the quantization of superconducting circuits, their design, fabrication and operation.

The first main result of this thesis, presented in Chpt. 2, is the implementation of a quantum Szilard engine. After a short theoretical introduction of the quantum Szilard engine, the latter is employed to hyperpolarize and thereby reveal the long-lived heretofore unknown TLS environment interacting weakly with the qubit.

As a second main result, a rigorous theoretical description of the underlying qubit TLS dynamics is presented in Chpt. 3. This comprises a derivation of the Solomon equations, adapted from the field of nuclear magnetic resonance and extended to an arbitrary number of TLSs, and a numerical modeling of the measured non-Poissonian quantum jumps statistics with the diffusive stochastic Schrödinger equation as well as with the Solomon equations.

Finally, the results are summarized in Chpt. 4 and an outlook with suggestions for future experiments is given.

2 Two-level system hyperpolarization with a quantum Szilard engine

In this chapter, the first main experimental results are presented, which led to the discovery of a long-lived two-level system environment of yet unknown physical origin that makes up the main loss mechanism of our granular aluminum fluxonium. This environment is made visible by the quantum Szilard engine. At the beginning of this chapter, the working principle of such an information-fueled thermodynamic engine is introduced. This chapter is adapted in parts from Spiecker et. al., Nature Physics 19, 1320–1325, 2023 (Ref. [21]).

2.1 The classical Szilard engine

To begin with, the classical Szilard engine as conceived by Szilárd [22] is briefly reviewed. The concept of the Szilard engine shows that information about thermodynamic fluctuations can be converted into negative entropies if one has in addition the requisite experimental capabilities. Fig. 2.1 depicts the working principle of the classical Szilard engine. The control of the Szilard engine can be imagined by an intelligent being or by a sophisticated autonomous working apparatus. The thermodynamic cycle begins with a classical particle in a box with unknown position and velocity. In the next step, a piston wall is inserted in the middle of the box, and a measurement is carried out, revealing weather the particle is trapped on the left or the right side. This bit of information is only converted into negative entropy when the entire system is now manipulated in such a way that the particle appears to be always on the correct side of the piston, for instance by rotating the box. In our experiment, we essentially flip the particle to the other side. With the weight attached to the piston, the box can now be attached to a heat reservoir at temperature T_r. In the subsequent isothermal expansion, the energy $\Delta W = -T_r \Delta S$ can be extracted from the reservoir, with

$$\Delta S = -k_B \ln 2 \tag{2.1}$$

being the entropy reduction of the particle in the box due to the intervention and k_B being the Boltzmann constant. Once the energy is extracted, the piston can be removed and the system returns to the starting point of the thermodynamic cycle. What we have achieved in total is that we gained energy by cooling down a reservoir.

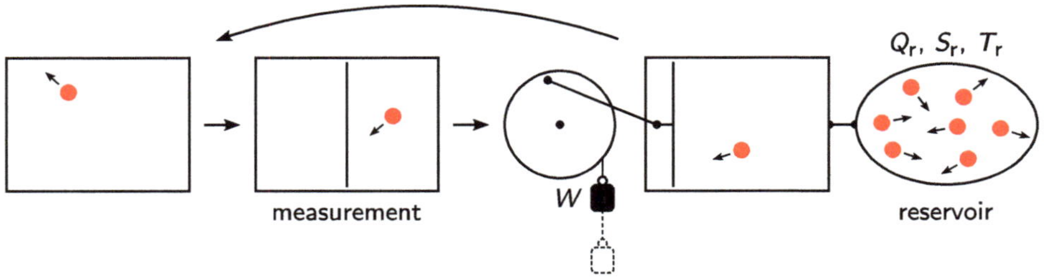

Figure 2.1: The classical Szilard engine. Thermodynamic cycle of the classical Szilard engine cooling a reservoir. Starting point is the particle in the empty box (left). After insertion of the piston wall and measurement of the particle position (middle) a weight is attached at the correct side of the piston. In the subsequent isothermal expansion the heat $\Delta Q_r = T_r \Delta S_r$ is extracted form the reservoir and converted into potential energy ΔW by lifting the weight (right). The cycle closes when the piston leaves the box.

This is clearly a violation of the second law of thermodynamics. When we consider a large number of these systems operated in parallel or sequentially, neither the measurement without manipulation nor the manipulation without access to the measurement information will extract a positive energy outcome on average. It is the conjunction of both ingredients that leads to the violation of the second law of thermodynamics. This conflict is usually solved by considering the entropy production in the intelligent being or the measurement apparatus that is caused by the erasure of the measured bit of information at the end of the cycle. According to Landauer's principle, at least the entropy $\Delta S_m \geq k_B T_a \ln 2$ is produced in the erasure process, with T_a being the temperature of the measurement apparatus [23–25]. For further discussions, see Ref. [26]. Regardless of whether or not information must be stored by the apparatus, one may instead assume that the second law is valid to allow for a theoretical comparison between different engines. In experiments, certainly much more energy is required for operating the system than can be obtained from the single particle.

2.2 The quantum Szilard engine

There are various ways in which the classical Szilard engine can be transferred to the quantum world. One possibility that is often discussed in the literature is to consider the quantum mechanical wave function of a quantum particle confined by the potential wells of the box [27, 28]. In this case, however, the operation and performance is slightly different compared to the classical Szilard engine since the isothermal insertion of the piston in form of a potential barrier can not be performed without additional work [27, 28]. In addition, the operation is not entirely quantum mechanical because the insertion and movement of the piston is described by classical variables, and hence, the extraction of energy results form a coupling of the quantum

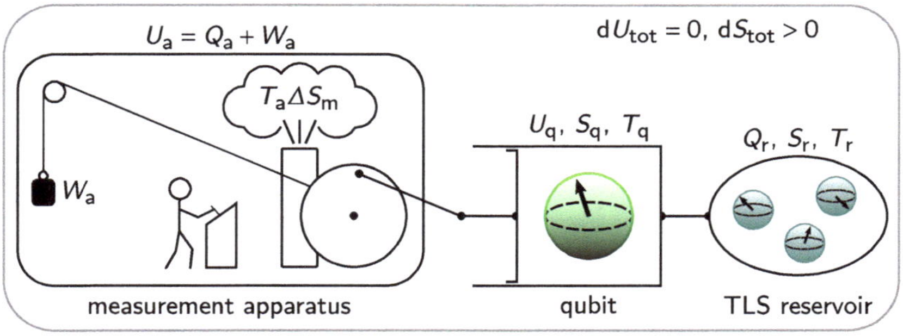

Figure 2.2: The quantum Szilard engine. Schematic of the entire thermodynamic system consisting of the measurement apparatus, the qubit and the TLS reservoir. For a closed system, the internal energy U_{tot} is constant, while the entropy S_{tot} can only increase irreversibly according to the second law of thermodynamics. The negative entropy production by the qubit preparation must therefore be compensated by an entropy production ΔS_m in the measurement apparatus, which is released at temperature T_a. As a consequence, performing the qubit preparation consumes the energy ΔW_a. The measurement and the feedback is performed by the intelligent being, as conceived by Szilard [22]. Reproduced from Ref. [21].

to the classical world. Instead, we will consider another implementation of quantum Szilard engines, depicted in Fig. 2.2, where the classical particle is replaced by a qubit [29–31]. This realization is very close to its classical counterpart, although three notable minor differences remain, as discussed below.

First of all, the qubit can be measured in its ground or excited state, which corresponds to the measurement of the classical particle to the left or right side of the piston. In thermal equilibrium, the probabilities of measuring the qubit in its ground or excited state are given by the Boltzmann statistics. These probabilities may also be emulated by the classical Szilard engine when the piston is inserted off-centered, meaning that the probabilities to measure a particle to the left or to the right of the piston are not equal.

Second, the qubit is brought in contact with a quantum mechanical heat reservoir, labeled TLS reservoir in Fig. 2.2. Consequently, the qubit and the reservoir are in general entangled. However, when a measurement of the qubit is performed, the entanglement with the environment is destroyed, resulting in a measurement back action on the environment, and the qubit is projected to one of its eigenstates. In the classical Szilard engine, no work is performed on the particle during the measurement. Interestingly, for the quantum Szilard engine, this is only true on average. The corresponding energy fluctuation as well as the measurement back action can be experimentally observed, as discussed in Sec. 3.6.3.

These two points show from a thermodynamic point of view, that immediately after the measurement the qubit can be interpreted as a classical particle.

The third point therefore constitutes the main difference, namely the extraction of energy: For the classical Szilard engine, an isothermal expansion converts the entropy in Eq. 2.1 into energy, no matter on which side the particle is found. In contrast, for the quantum Szilard engine, energy can only be extracted when the qubit was projected to the excited state, for instance by a stimulated emission. This means that only the average qubit energy is extracted, which is less than for the classical Szilard engine even if the position of the piston is adjusted to the populations of the qubit.

2.2.1 Measurement and entropy

Completing Szilard's thought experiment, we imagine the measurement apparatus with an energy storage W_a that provides the energy required for the measurement process and a thermal bath at temperature T_a, where the negative entropy from the projected qubit can be compensated. Clearly, as we consider the whole system to be closed, the inner energy U_{tot} must be conserved and, according to the second law of thermodynamics, the entropy S_{tot} can only increase irreversibly (Fig. 2.2).

For the quantitative discussion of the Szilard engine, we will consider a generalized qubit system with potentially degenerate energy levels. This allows us to show that the measurement and reset of the qubit can produce different amounts of entropy, which has to be considered when comparing the efficiencies of various information powered engines. The energy level structure is depicted in Fig. 2.3. For $n = m = 1$ we have the experimental situation of the qubit.

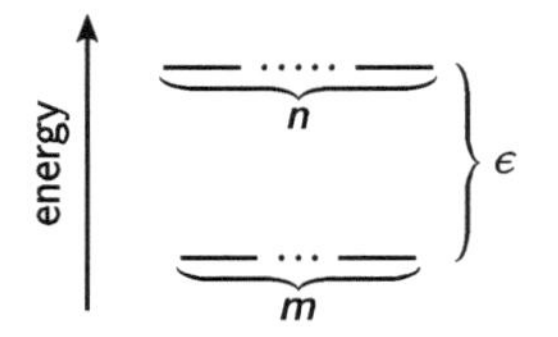

Figure 2.3: Generalized qubit system with degenerate energy level manifolds and energy difference ϵ.

In thermal equilibrium, the qubit system is measured with the probabilities

$$p = \frac{n}{n + me^{\beta\epsilon}} \quad \text{and} \quad q = \frac{me^{\beta\epsilon}}{n + me^{\beta\epsilon}} \tag{2.2}$$

in the excited and ground state manifold, respectively, with the qubit energy ϵ and $\beta = 1/k_B T_r$ defined by the temperature T_r of the reservoir. The internal energy of the qubit is simply $U_q = p\epsilon$. Furthermore, considering the statistical ensemble, the entropy amounts to

$$\frac{S_q}{k_B} = -\frac{n}{n + me^{\beta\epsilon}} \ln\left(\frac{1}{n + me^{\beta\epsilon}}\right) - \frac{me^{\beta\epsilon}}{n + me^{\beta\epsilon}} \ln\left(\frac{e^{\beta\epsilon}}{n + me^{\beta\epsilon}}\right) \tag{2.3}$$

$$= \beta U_q - \beta\epsilon + \ln\left(n + me^{\beta\epsilon}\right). \tag{2.4}$$

Next, we first focus on the less entropy demanding measurement that projects the qubit system onto its excitation manifolds, without resolving the individual levels.

If the experiments were post-selected according to the measurement outcome, the entropies

$$\frac{S_{q,|g\rangle}}{k_B} = \ln m, \quad \text{and} \quad \frac{S_{q,|e\rangle}}{k_B} = \ln n \tag{2.5}$$

would be obtained for a projection to the ground or excited energy manifold, respectively.

When taking the statistical average over all experiments, the measurement must be followed by a conditional reset to the target manifold in case the system was projected to the opposite manifold. However, if the manifolds are of different size $n \neq m$, a full reset is not possible, as it would mean to destroy the difference in entropy. In other words, since the reset is a unitary operation, both manifolds must have the same dimensionality, and hence $m = n$. With this constriction the average entropy reduction after a conditional reset is simply

$$\frac{\Delta S_q}{k_B} = p \ln p + q \ln q, \tag{2.6}$$

which is always negative. Once again, the information of the measurement alone is not equivalent to entropy reduction. Negative entropy is only produced in conjunction with the ability to react to the thermodynamic fluctuations, i.e. information is gained in terms of Shannon's entropy. The maximum average entropy reduction of $\Delta S/k_B = -\ln 2$ is attained for equal population in the manifolds, which implies an infinite temperature or a vanishing energy level splitting.

Now, we turn to a more powerful measurement that can distinguish between all $m + n$ states. After such a measurement, the system is in a pure state. It produces a maximum average entropy reduction of $\Delta S/k_B = -\ln(n+m)$ in the limit of $\beta\epsilon \to \infty$, in accordance with the previously mentioned limit for the qubit. This is also the reason why the quantum Szilard engine in Ref. [32], which operates on several bosonic particles, can exceed the work output $\Delta W = k_B T_r \ln 2$. If their Szilard engine had been equipped with gate operations that could shuffle bosons from one side of the piston to the other, even larger output powers would be achievable. In fact, the same entropy reduction may be interpreted and obtained by several binary measurements $\Delta S/k_B = -\ln 2 \log_2(n+m) = -\log(n+m)$. We see here that for a fair comparison of engine efficiencies, the energy cost of a measurement must be taken into account, which, following Szilard's reasoning, results from the assumption that the second law of thermodynamics should remain valid.

2.2.2 Efficiency

The Szilard engine can be used both to cool and to heat the reservoir. The heat pump and refrigeration cycles are depicted in Fig. 2.4 for a qubit with thermal excitation as in the experiment.

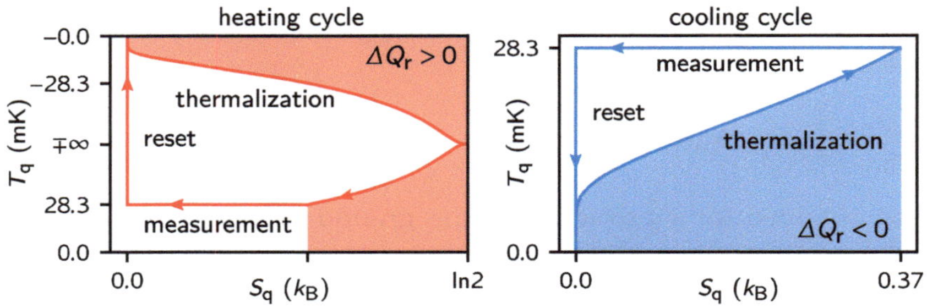

Figure 2.4: Quantum Szilard engine heating and cooling cycle. Temperature-entropy diagram of the heating (left panel) and cooling cycle (right panel) of the quantum Szilard engine. Note the negative temperatures for the heating cycle indicating the population inversion of the qubit. Before the measurement the qubit is in thermal equilibrium with the reservoir. The colored areas correspond to the heat transferred with the reservoir. They amount to $\Delta Q_r = \epsilon - U_q$ and $\Delta Q_r = -U_q$ for heating and cooling, respectively. Reproduced and adapted from Ref. [21].

Since the measurement apparatus is considered to be a thermodynamic engine, the entropy reduction in the qubit system has to be compensated by an entropy production $\Delta S_m \geq -\Delta S_q$ by the measurement apparatus in order for the second law of thermodynamics to remain valid. Consequently, the apparatus must be connected to a heat bath into which it can unload the entropy. When this bath is at the temperature T_a, the measurement requires the work $\Delta W_m = -T_a \Delta S_m$. Furthermore, one can argue that the measurement process should not depend on the temperature of the qubit system. Thus, for the binary measurement $\Delta S_m \geq k_B T_a \ln 2$ must hold, as was first conjectured by Szilárd [22]. In the following we will drop this assumption and use the exact entropy reduction, since it allows for a simple comparison with the theoretical maximum performance given by the Carnot cycle.

Cooling cycle

In order to cool the reservoir, the qubit is reset to its ground state. From the thermalization process after the reset it follows $\Delta U_q = U_q > 0$. With this definition it follows for the three steps:

1. **Measurement:** The measurement requires the work $-T_a \Delta S_m$.
2. **Reset:** From the reset of the qubit to the ground state one can in principle extract in average the work ΔU_q, which may be used to lift up again the weight of the measurement apparatus.
3. **Thermalization:** The reservoir is cooled down by $\Delta Q_r = -\Delta U_q < 0$. Here, the TLS reservoir is assumed large enough so that its temperature T_r stays approximately constant.

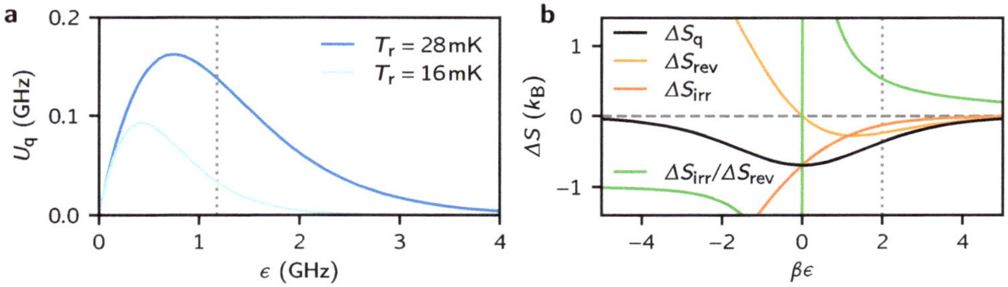

Figure 2.5: Energy and entropies for the cooling cycle. a Average qubit energy as a function of the qubit energy splitting. The dark blue curve corresponds to the measured thermal qubit temperature, while the light blue curve corresponds to the temperature that is reached after cooling the TLSs. **b** Various entropies for the cooling cycle as defined in the text as a function of the normalized inverse temperature $\beta\epsilon = \epsilon/k_B T_r$ of the reservoir. The dotted gray lines in both panels indicate the experimental situation.

The coefficient of performance (COP) is defined as the ratio of extracted heat and amount of energy used, and hence

$$\mathrm{COP} = \frac{|\Delta Q_r|}{-\Delta W_{tot}} = \frac{\Delta U_q}{T_a \Delta S_m - \Delta U_q} \tag{2.7}$$

Since the reservoir did not change its temperature, a reversible entropy can be defined via $\Delta Q_r = T_r \Delta S_{rev} < 0$, which is also valid when the temperature in the bath is negative, in which case ΔS_{rev} becomes positive. We may then write, assuming $\Delta S_m = -\Delta S_q$ holds

$$\mathrm{COP} = \frac{-T_r \Delta S_{rev}}{-T_a(\Delta S_q - \Delta S_{rev} + \Delta S_{rev}) + T_r \Delta S_{rev}} = \frac{T_r}{T_a - T_r + T_a \frac{\Delta S_{irr}}{\Delta S_{rev}}}, \tag{2.8}$$

where we defined the irreversible entropy reduction $\Delta S_{irr} = \Delta S_q - \Delta S_{rev} < 0$ that emerges from the measurement process. From Eqs. 2.4 and using that $m = n$, it follows

$$\frac{\Delta S_{irr}}{k_B} = \beta\epsilon - \ln\left(1 + e^{\beta\epsilon}\right) = -\ln\left(1 + e^{-\beta\epsilon}\right) < 0. \tag{2.9}$$

For the entropy ratio one obtains

$$\frac{\Delta S_{irr}}{\Delta S_{rev}} = \frac{1 + e^{\beta\epsilon}}{\beta\epsilon} \ln\left(1 + e^{-\beta\epsilon}\right) = \frac{1}{\beta\epsilon} + \frac{1}{2\beta\epsilon e^{\beta\epsilon}} + \dots \quad \text{for} \quad \beta\epsilon \to \infty, \tag{2.10}$$

showing that the Szilard engine refrigerator can reach the maximum theoretical efficiency in the limit of vanishing temperature or infinite qubit energy. However, as is typical for purely reversible processes, the cooling power given by $\Delta U_q = \beta\epsilon/(1 + e^{\beta\epsilon})$

vanishes rapidly. The temperature dependencies of the average qubit energy as well as of the various entropies are shown in Fig. 2.5.

Note that the COP remains correct even if the temperatures of the reservoir and the measurement apparatus are negative. A negative COP simply means that energy is extracted as a byproduct from the inverted reservoirs.

Heating cycle

In order to heat the reservoir the qubit must be reset to its excited state. From the thermalization process after the reset, it follows $\Delta U_q = U_q - \epsilon < 0$. With this definition it follows for the three steps:

1. **Measurement:** The measurement requires the work $-T_a \Delta S_m$.
2. **Reset:** The reset of the qubit to the excited state requires the additional work ΔU_q on average.
3. **Thermalization:** The reservoir is heated by the amount $\Delta Q_r = -\Delta U_q > 0$. As before, the TLS reservoir is assumed large enough so that its temperature T_r stays approximately constant.

Analogously, it holds

$$\mathrm{COP} = \frac{|\Delta Q_r|}{-\Delta W_{tot}} = \frac{\epsilon - U_q}{T_a \Delta S_m + \epsilon - U_q} = \frac{T_r}{T_r - T_a + T_a \frac{\Delta S_{irr}}{-\Delta S_{rev}}}. \tag{2.11}$$

Here, we have $\Delta S_{rev}/k_B = \beta\epsilon - \beta U_q$ for the reversible entropy and the irreversible entropy amounts to $\Delta S_{irr}/k_B = -\ln(1 + e^{\beta\epsilon}) < 0$. For the entropy ratio, we therefore obtain

$$\frac{\Delta S_{irr}}{-\Delta S_{rev}} = \frac{1 + e^{-\beta\epsilon}}{\beta\epsilon} \ln\left(1 + e^{\beta\epsilon}\right), \tag{2.12}$$

which, in contrast to Eq. 2.10, vanishes for negative temperatures.

2.3 The granular aluminum fluxonium

After having introduced the working principle of the Szilard engine, the remaining sections of this chapter deal with the experimental realization and its application to uncover hidden environments. The quantum Szilard engine is realized with a superconducting fluxonium qubit, which uses the high kinetic inductance of granular aluminum for its superinductor.

In the following, the granular aluminum fluxonium is briefly presented, while further in-depth discussions on the design and implementation are provided in Part II of this

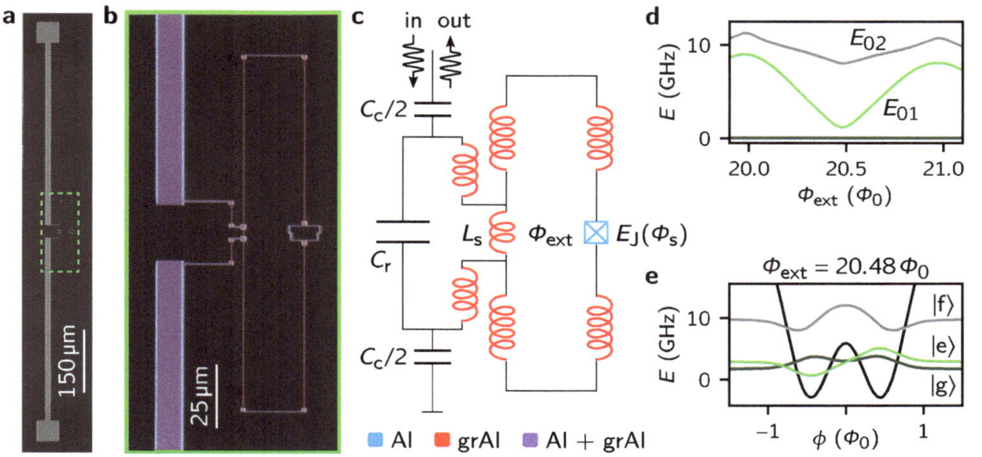

Figure 2.6: The granular aluminum fluxonium with a SQUID-junction and readout resonator. a, b False-colored optical image of the granular aluminum fluxonium and its readout antenna. The device is fabricated in a single three-angle evaporation lithography step. The first two aluminum layers (blue) are used to implement the conventional Al/AlO_x/Al-junction and form the antenna. The granular aluminum zero angle evaporation implements all the inductances. The granular aluminum wires have a width of ~ 200 nm (red). The regions where the granular aluminum is shunted by aluminum underneath are colored in violet. **c** Equivalent circuit of the resonator fluxonium system. The antenna couples via the capacitance C_c to the microwave waveguide for readout and qubit control. More details on the circuit and parameters are provided in Sec. 6.2 and Tab. 6.1. **d** Energy spectrum of the fluxonium as a function of the applied external flux Φ_{ext} around the half flux sweet spot at which the experiments were performed. **e** Fluxonium potential and representation of the first three wave functions at the half flux sweet spot shown in panel d. The fluxonium parameters are $L_q = 234$ nH, $C_q = 7.0$ fF and at this flux position $E_J = 5.9$ GHz. Reproduced and adapted from Ref. [18].

thesis. The fluxonium consists of a Josephson junction that is shunted by a so-called superinductance [15]. Typically, these superinductors are implemented by Josephson junction arrays. For the granular aluminum fluxonium, the junction array is simply replaced by a granular aluminum wire. Initially, the main motivation was to greatly reduce the number of modes in the wire (Sec. 5.2) and thereby potentially improve the qubit readout. Concomitant, the coherence properties of granular aluminum can be studied, which is illustrated by this work.

A granular aluminum fluxonium inductively coupled to its readout antenna, is depicted in Fig. 2.6a, b. The circuit is fabricated in a single lithography step. A detailed description of the fabrication is given in Sec. 7.1. The junction of the fluxonium features a superconducting quantum interference device (SQUID), which allows the Josephson energy of the fluxonium to be tuned with an externally applied magnetic field. In this way, the fluxonium-resonator system can be brought in a regime with excellent readout conditions, as we reported in [33]. The equivalent circuit of the fluxonium and readout resonator system is shown in Fig. 2.6c. Quantization of this circuit yields the fluxonium spectrum and information about the dispersive frequency shift of the resonator as a function of the fluxonium state. A general introduction to circuit quantization is given in Chpt. 5. The quantization of the bare fluxonium as well as of the fluxonium together with its readout resonator is detailed in Sec. 5.3 and Sec. 6.2, respectively. Fig. 2.6d depicts the fluxonium spectrum with respect to the ground state energy around the half flux sweet spot, at which most of the measurements were carried out. At this half flux position, the fluxonium spectrum forms a well defined qubit system with qubit frequency $f_\mathrm{q} = 1.2\,\mathrm{GHz}$ separated by 6.8 GHz from the higher levels. The corresponding wave functions are plotted in Fig. 2.6e. An enlarged view of the measured and calculated spectrum is shown in Sec. 5.3.2. The device is placed in a microwave waveguide with an extra port to drive the qubit. For furter details on the microwave setup see Sec. 7.3.

2.3.1 Quantum non-demolishing readout and active reset

The implementation of the quantum Szilard engine ideally requires a textbook quantum mechanical measurement of the qubit and depending on the measurement outcome an active decision and control to reset the qubit to its target state. Performing a textbook quantum mechanical measurement means that the qubit is projected to the corresponding eigenstate of the measurement outcome, where it can then be measured again with absolute certainty. As was discussed in more detail in Sec. 2.2, only both ingredients, measurement and control, produce a pure qubit state on average.

The non-destructive measurement is by no means trivial to achieve and led to the concept of quantum non-demolishing (QND) measurements. Nowadays, the QND readout of superconducting qubits is achieved by measuring the dispersive frequency shift of the qubit's readout cavity [34, 35]. Further insights on how to design the dispersive readout are provided in Chpt. 6.

The dispersive readout, as typically implemented, is only approximately QND. With increasing numbers of circulating photons in the readout resonator detrimental demolishing effects are usually encountered. These demolishing effects can also be observed in our device. Fortunately, they occur at different flux values than those used in the experiment, as can be seen in Fig. 6.3b and Fig. 6.6. The origin of this non-QND behavior has puzzled the community for almost two decades [36–39], and it is only recently that non-perturbative methods can predict these non-QND effects almost quantitatively at large photon numbers [40]. The fact that one is essentially forced to use low photon numbers led to the development of parametric amplifiers placed downstream, as close as possible to the sample [41–44].

Surprisingly, we find that most of our granular aluminum fluxoniums can be read out at half flux using large photon numbers. For the fluxonium presented here, an active reset fidelity without using a parametric amplifier of 99 % and 93 % for the ground and excited state, respectively, could be achieved with $\bar{n} = 74$ photons and an

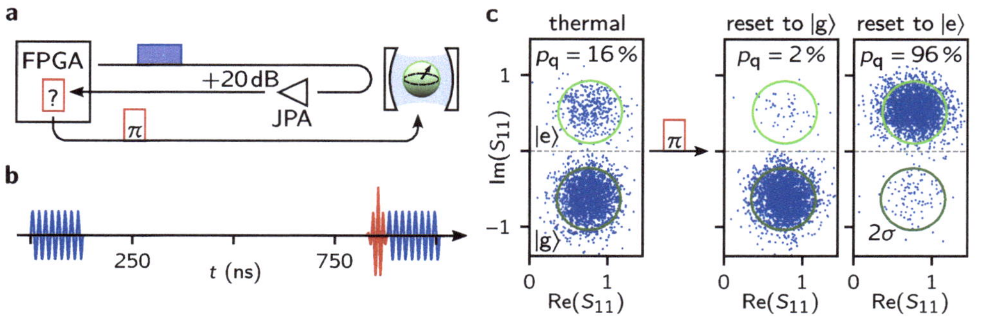

Figure 2.7: Qubit readout and active reset. **a** Schematic drawing of the microwave setup for measuring the readout cavity and controlling the qubit. The experiments are orchestrated by the field programmable gate array (FPGA) controller from Quantum Machines, with an internal real-time feedback latency of ~ 200 ns. **b** Pulse sequence and timings of the active reset. The readout pulse (blue) that excites the resonator is 128 ns long. The integration of the resonator ring-down, with decay time τ = 280 ns, occurs after 400 ns, once the distorted input signal has left and the qubit is projected. In case the qubit has to be reset, a π-pulse with a Gaussian envelope of σ = 10 ns is played in a 48 ns window (red). The readout pulses are separated by 880 ns. **c** Scatter plot of the complex reflection coefficient S_{11} of the readout signal for the qubit in equilibrium (left panel) and after qubit preparation in $|g\rangle$ and $|e\rangle$ (right panels). Adapted from Ref. [21].

integration time of about half a microsecond [33]. The reason for the photon resilience is not entirely clear, but could have its origin in the absence of low-lying modes present in junction array fluxoniums, as discussed in Sec. 5.2. Nevertheless, the parametric amplifier, tuned to 20 dB of gain, is of great value for the operation of the quantum Szilard engine, as it allows for a significant reduction of $\bar{n}$, thereby allowing the qubit to interact freely with its environment in between the measurements. In addition, in future experiments one could use a phase controlled pulse to empty the cavity after the readout [45, 46].

A schematic drawing of the experiment with the microwave electronics providing the real time capabilities is shown in Fig. 2.7b. A complete drawing of the microwave setup can be found in Sec. 7.3. The intermediate frequency pulse sequence for the active reset that is generated by the field programmable gate array (FPGA) electronics is depicted in Fig. 2.7c. The integration of the resonator ring-down occurs relatively late after 400 ns once the distorted readout signal has left the system. The internal latency of the controller to set up the conditional π-pulse and the subsequent measurement pulse is about 200 ns. Fig. 2.7c depicts the active reset from thermal equilibrium to $|g\rangle$ and $|e\rangle$ as used for the quantum Szilard experiment. As will be shown later in Fig. 2.16a, the presence of a long-lived environment, which is the main loss mechanism for the qubit, allows improving the active reset by polarizing the environment beforehand. In this way, an active reset of 99.8 % for $|g\rangle$ and 98.3 % for $|e\rangle$ could be achieved.

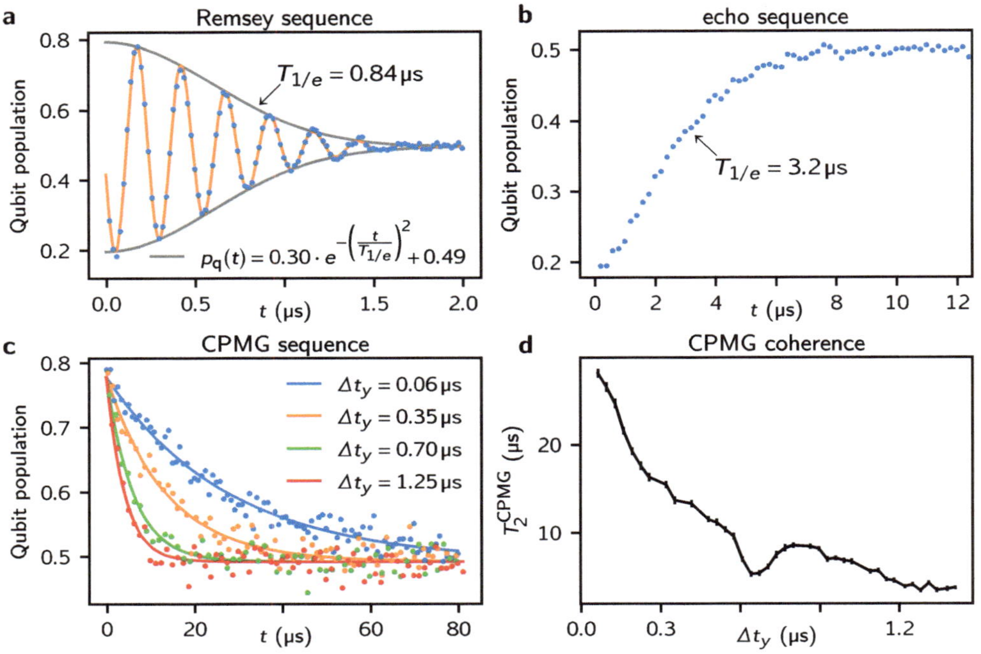

Figure 2.8: Coherence measurements of the qubit. **a** Measured Ramsey decay with Gaussian envelope (blue points are measured and solid lines are fits). **b** Echo decay, which contains both exponential and Gaussian components. The arrow indicates the $1/e$-time. **c** Selection of measured CPMG decays (points) with exponential fits (solid lines). **d** Coherence times as a function of the y-pulse repetition time in the CPMG sequence. The coherence times and the corresponding standard deviation errors are obtained from the exponential fits (panel c). The lines connecting the markers with error bars are guides to the eye. Reproduced from Ref. [21].

2.3.2 Coherence of the SQUID-fluxonium

Before the qubit relaxation is discussed in detail in the following sections, the coherence properties of the SQUID-fluxonium are briefly presented. In Fig. 2.8 various coherence measurements of the fluxonium are depicted. The relatively low coherence compared to the qubit's relaxation time $T_1 = 21.5\,\mu s$ is most likely caused by local flux noise which, leads to a fluctuating Josephson energy of the SQUID-junction. Since the flux-dependent Josephson energy is operated at a point with non-zero slope (s. Fig. 5.15), the qubit frequency is in first order sensitive to local flux noise, even if it is at a sweet spot with regards to the global flux. In contrast, for one of the standard granular aluminum fluxoniums with only one junction that have been fabricated on the same wafer, a coherence time of $T_2 = 28\,\mu s$ was measured, while for the echo sequence a coherence time of $T_2 = 46\,\mu s$ was achieved, close to the limit given by the qubit's relaxation time [18].

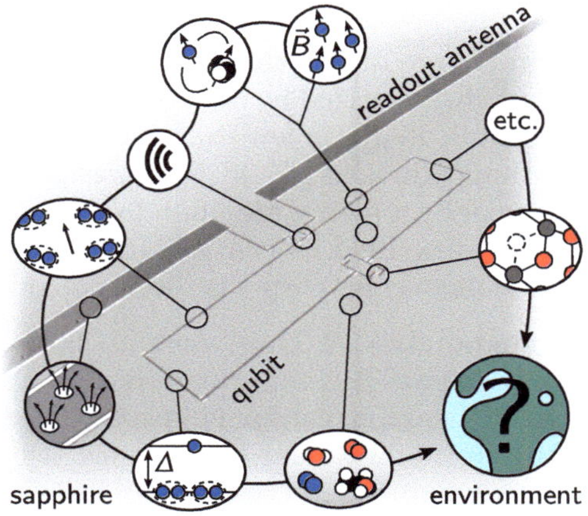

Figure 2.9: Superconducting qubit environments. Schematic drawing of the fluxonium qubit and the rich environment typical for superconducting circuits. It includes counterclockwise: free electronic spins that may be Zeeman split by an external magnetic field [47, 48] or via the hyperfine interaction [49], radiation loss into the readout and qubit drive ports [50] or into spurious modes including phonons [51], Shiba spins [52], trapped vortices [53, 54], quasiparticles [55], adsorbed molecules on the surface [56], and dielectric TLSs [57]. Reproduced from Ref. [21].

The Gaussian envelope for the Ramsey sequence in Fig. 2.8a reveals a significant contribution of slow noise. Consequently, the coherence can be drastically increased with the number of refocusing pulses (Fig. 2.8b,c). The decay of the Carr-Purcell-Meiboom-Gill (CPMG) sequence is well described by an exponential function with a maximum coherence exceeding the qubit's relaxation time (Fig. 2.8d). Since the coherence is closely related to the qubit's frequency noise, the dip visible at $f = 1/2\Delta t_y = 0.77\,\text{MHz}$ indicates an excess noise source. While the frequency noise is expected to come from flux noise, it is possible that the TLS environment also contributes to this frequency noise and overall decoherence behavior.

2.4 Superconducting qubit environments

The innate complexity of solid-state physics exposes superconducting quantum circuits to interactions with uncontrolled degrees of freedom, which degrade their coherence. For this reason, superconducting qubits are relatively short-lived compared to other quantum platforms such as nitrogen vacancy centers [58] and donor spins in silicon [59]. The complex environment that can be observed in superconducting circuits is depicted in Fig. 2.9. In order to improve the qubit life time it is important to identify the main loss mechanism of the qubit.

While some of these environments stand out due to their very clear signature, for instance an increased loss rate at the hyperfine transition of hydrogen [49] or frequency fluctuations in charge qubits due to a fluctuating quasiparticle background [60], others are best characterized by directly manipulating the environment of the superconducting circuit. To give a few examples, free electronic spins can be tuned in resonance with magnetic field [47, 48, 61], quasiparticles can be generated by strong saturation pulses or by ionizing radiation [62, 63], and dielectric TLSs can be tuned with strain and electric field [57, 64, 65]. In the case of strongly coupled TLSs one may even operate them coherently via the superconducting qubit [64–67].

Speaking now of the general class of TLSs, such as spins, dielectric TLSs, and potentially also trapped quasiparticles, the situation changes drastically when the TLSs are weakly coupled to the qubit, cannot be resolved spectrally, and are short-lived compared to the qubit lifetime. These TLSs may still be investigated with harmonic oscillators and by varying the TLS saturation [68–70]. However, as we will show in Chpt. 3, for a qubit these TLSs provide a Born-Markovian environment with a relaxation that is independent of the TLS populations. In addition, due to the short lifetime, the TLS population remains unchanged by qubit operations. As the last strategy, one may vary the qubit coupling to capacitive, inductive, and other lossy environments and thereby identify the main loss mechanism [71].

If the TLSs have a similar or even longer lifetime than the qubit, direct manipulation and observation of the TLS environment is possible again. In this case, one may either use simple saturation pulses (Sec. 2.6) or, more effectively, hyperpolarize the TLSs with the quantum Szilard engine protocol, similar to the hyperpolarization protocols available for nuclear magnetic spins [72, 73]. The beauty of the quantum Szilard engine approach is its independence of the physical origin of the TLSs, which allows to uncover heretofore hidden environments.

2.5 Origins of non-exponential relaxation

There exist many possible reasons for observing non-exponential qubit relaxation. Often the non-exponential relaxation is caused by a fluctuating relaxation time. For instance, a steep relaxation at the very beginning is likely to originate from an occasional quasiparticle burst produced by cosmic or ambient radioactive radiation [63, 74, 75]. A bi-exponential relaxation can hint at a nearby toggling dielectric TLS [76] or a fluctuating quasiparticle background [71]. Assuming the latter scenario, the relaxation of the qubit population p_q from the excited state can be described by

$$p_\mathrm{q}(t) = (1 - p_\mathrm{th}) \cdot \left(q \cdot e^{-t/T_\mathrm{q}} + (1 - q)\right) \cdot e^{-t/T_\mathrm{r}} + p_\mathrm{th}, \tag{2.13}$$

with the probability q that a quasiparticle is present giving rise to an additional decay rate defined by T_q, whereas in the absence of the quasiparticle the qubit relaxes with the decay time T_r to the thermal equilibrium population p_th.

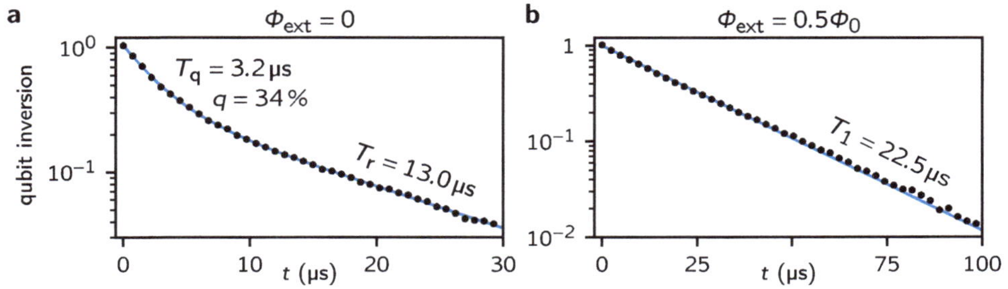

Figure 2.10: Flux dependent relaxation of the single junction fluxonium. **a** Bi-exponential relaxation at zero flux. The continuous line corresponds to the model in Eq. 2.13. **b** Exponential relaxation at half flux. Reproduced from Ref. [18].

In principle, one can generalize the relaxation to an arbitrary number of exponential functions and corresponding probabilities. For quasiparticles, one often assumes a Poisson distribution $P_\lambda(n)$ and a relaxation that increases linearly with the number n of quasiparticles [71]. This distribution is rather mathematically motivated, as it allows the relaxation to be expressed again in closed form. It holds [71]

$$p_q(t) = (1 - p_{th}) \cdot e^{-\lambda(1-e^{-t/T_q})} \cdot e^{-t/T_r} + p_{th}. \tag{2.14}$$

From a physical point of view, the Poisson distribution is rather unlikely because of the quadratically increasing recombination probability of the quasiparticles [62, 74]. Especially in light of quasiparticle-induced qubit losses at the tiny fluxonium junction [71], the presence of several quasiparticles seems implausible. In Refs. [71, 77], and in Fig. 2.10 adapted from Ref. [18], a bi-exponential relaxation perfectly describes the experimental findings. Since flux qubits biased at half flux are insensitive to quasiparticle loss at the junction due to the destructive quasiparticle interference [55, 71], the presence and absence of a non-exponential relaxation at zero and half flux (Fig. 2.10) hints at the presence of a quasiparticles.

In order to determine whether a given set of relaxation curves can be attributed to the same fluctuating qubit relaxation, one can subtract a single thermal equilibrium population from all relaxation curves and then normalize the curves by the initial amplitudes. If the curves do not match, the underlying statistics may have changed, or the relaxation dynamics could be caused by a memory effect in the environment, as will be shown in the next section.

2.6 Heating with π-pulses

A simple experiment to find out whether the qubit is coupled to a finite size long-lived environment is to repeatedly excite the qubit by a sequence of π-pulses [78].

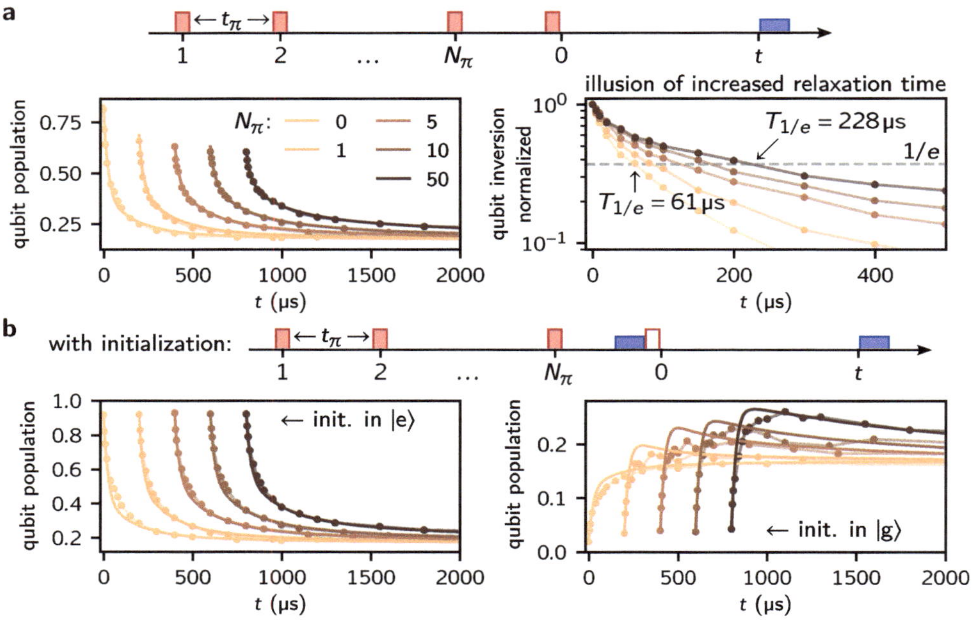

Figure 2.11: Heating without active feedback. a Top panel: schematic of the pulse sequence consisting of N_π repeated π-pulses spaced by $t_\pi = 100\,\mu s$, followed by a standard free decay experiment. The value for t_π is chosen larger than the intrinsic qubit decay time but smaller than the relaxation of the environment in order to heat the environment. Left panel: Free decay of the qubit for various N_π. The curves are shifted horizontally for visibility. Right panel: Measured relaxation curves taken from the left panel, normalized and plotted in log-scale. For our device, the ostensibly increased relaxation time is an illusion, and it is explained by the increased environmental temperature which in turn heats the qubit, as demonstrated in panel b (right panel). Consequently, this heating of the environment forbids to compare scaled and shifted non-exponential relaxation curves. **b** Top panel: schematic of the pulse sequence similar to the one in panel a, except that the free decay is measured following qubit initialization in $|g\rangle$ or $|e\rangle$. The left panel measurements after initialization in $|e\rangle$ appear similar to the corresponding ones in panel a, while, strikingly, after initialization to $|g\rangle$ (right panel) non-monotonic evolutions of the qubit population are observed due to the heat stored in the environment. The solid lines are simultaneous fits using the theoretical model of Eqs. 2.22 and 2.22 including the π-pulse sequence on the qubit. Reproduced from Ref. [21].

One should of course make sure that the altered relaxation and temperature is not caused by a trivial heating of the qubit environment, for instance a heating of the microwave setup from the π-pulses. A good control experiment is to replace the π-pulses with 2π-pulses to make sure it is indeed the qubit excitation that causes the different behavior.

A sequence (Fig. 2.11a, top) that does not require active reset capabilities was introduced in Ref. [78] and applied to flux qubits. The π-pulses are separated by $t_\pi = 100\,\mu s$, chosen to be longer than the intrinsic qubit decay time but shorter than the relaxation of the environment, in order to heat the environment. The relaxation dynamics of the SQUID-fluxonium depending on the number N_π of π-pulses is shown in Fig. 2.11a (left). Since the curves do not match after the normalization, as described in the previous section, the environment of the qubit must have changed. This, however, does not allow to claim an increased relaxation time of the qubit. For our qubit, the seemingly increased T_1-time is simply due to the heated environment, as will be shown using an active reset sequence.

Instead of measuring the qubit relaxation with the remaining qubit polarization, in the second sequence (Fig. 2.11b, top) the qubit is actively reset to $|e\rangle$ or $|g\rangle$ from where the free relaxation to thermal equilibrium is measured. The relaxation is shown in Fig. 2.11b (left and right panels). Of particular importance is the qubit initialization to the ground state. In the subsequent evolution the qubit heats up and overshoots its thermal population p_{th} proving an increased temperature of the qubit environment. The theoretical model shown in Fig. 2.11 follows from the integration of the Solomon equations (Eqs. 2.22 and 2.23) and includes the π-pulse sequence.

2.7 Hyperpolarization with the quantum Szilard engine

To enhance the manipulation of the qubit environment compared to the π-pulse sequence presented in the previous section, a quantum Szilard engine can be implemented using the qubit and its active feedback control. Alternatively, one can also use an autonomous feedback [72]. For instance, the fast flux tunable fluxonium presented in Sec. 7.1 can be tuned in resonance with its readout resonator to quickly dump the qubit excitation into the transmission line. Back at the operating point, the qubit can be inverted if the environment is to be heated. In fact, this protocol implements the quantum Otto engine [79].

As will be seen later, the Szilard engine hyperpolarizes the qubit environment and can even produce negative temperatures in the environment. The hyperpolarized environment reveals that the qubit is weakly coupled to a TLS environment of unknown origin, which relaxes over tens of milliseconds. Conversely, this so-far undiscovered environment can now be identified as the dominant loss mechanism of our qubit, and one may dread that similarly acting environments are ubiquitous in superconducting hardware.

In order to efficiently manipulate the TLS environment, the TLS must ideally provide the main loss mechanism for the qubit while in turn being approximately lossless, as schematized in Fig. 2.12a. The polarization sequence for heating or cooling the TLSs is depicted in Fig. 2.12b. Each cycle of the Szilard engine consists of a qubit preparation, which is followed by a cross relaxation between the qubit and the TLSs. In this way, photons that leave the qubit-TLS system are immediately detected

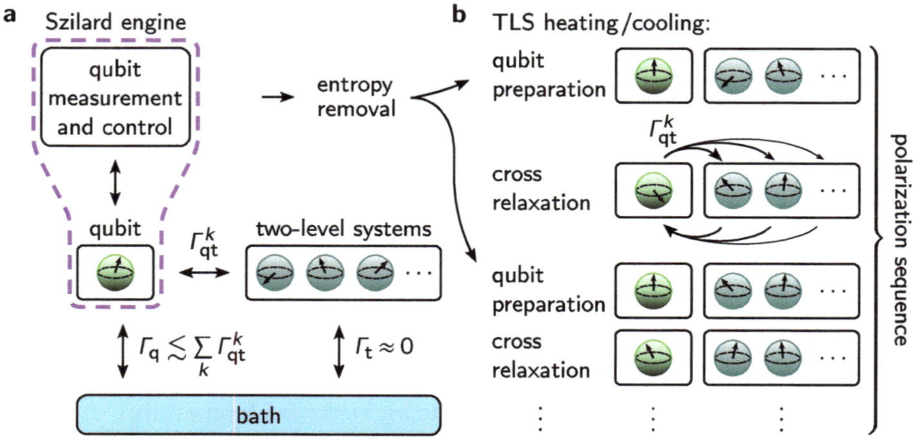

Figure 2.12: Working principle of the Szilard engine. a As will be shown in Sec. 2.8, the qubit environment can be modeled as a collection of polarizable TLSs and a global bath responsible for the so-called "intrinsic loss" of both qubit (Γ_q) and TLSs (Γ_t). The TLSs act as heat reservoir, because they provide the main relaxation channel for the qubit ($\Gamma_q \lesssim \sum_k \Gamma_{qt}^k$) while being approximately lossless ($\Gamma_t \approx 0$). **b** Schematic illustration of the qubit and TLS populations during the polarization sequence. Each cycle of the Szilard engine consists of a qubit preparation followed by the cross relaxation between the qubit and the TLSs. After each cycle the polarization of the TLSs increases. Reproduced from Ref. [21].

and, depending on the desired polarization, removed or added to the system again. Consequently, the TLS polarization follows the actively stabilized qubit state, and the amount of heat in the TLS environment varies with the operation time of the Szilard engine.

2.7.1 Free decay measurements

The experimental workflow to investigate the TLS environment is depicted in Fig. 2.13. The experiment begins with the TLS polarization sequence by stabilizing the qubit to either $|g\rangle$ or $|e\rangle$ using N active feedback preparations. This sequence implements the quantum Szilard engine as proposed in Fig. 2.12b. The active feedback preparations are separated by $t_{rep} = 2\,\mu s$, which is much shorter than the qubit lifetime of $T_1 = 21.5\,\mu s$ to reliably detect each photon passing the qubit, but also longer than the 140 ns ring-down time of the ~ 20 resonator photons so that most of the time the qubit and the TLSs can freely interact with each other in between the measurements. Having realized in Fig. 2.11b the importance of measuring the qubit relaxation from $|g\rangle$ or $|e\rangle$, the polarization is followed by a qubit initialization.

The qubit relaxation is shown in Fig. 2.14a after heating the TLSs for various times $N \cdot t_{rep}$. Note the logarithmic x-axis from 300 µs onwards, which is necessary to

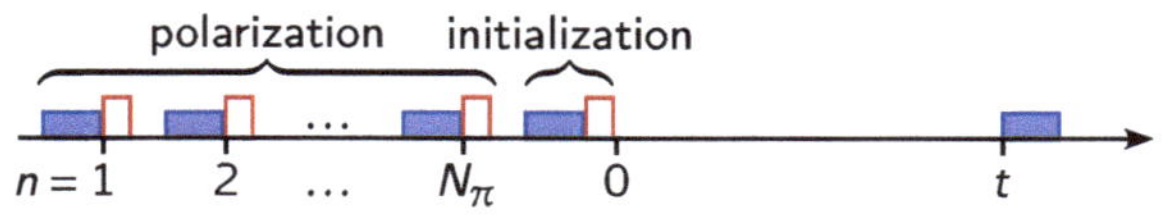

Figure 2.13: Control sequence implementing the Szilard engine and the free decay measurement. The active feedback preparations implementing the Szilard engine are separated by 2 µs, much shorter than the qubit's relaxation time T_1 = 21.5 µs, but longer than the resonator is populated with photons. The polarization sequence is followed by a qubit initialization to $|g\rangle$ or $|e\rangle$ to measure the free decay of the qubit. Adapted from Ref. [21].

depict the slow relaxation visible over the course of 50 ms. For the relaxation with $N = 10^4$ and qubit initialization to $|g\rangle$, it can be seen that the qubit rapidly heats up and reaches a population inversion of $\sim$ 60 %.

Due to the long relaxation time, measuring the entire data set with 10000 single shot qubit measurements for each of the 52 time points would take more than three days if the system was always given 50 ms to relax to thermal equilibrium. Here, the measurement time was reduced to $\sim$ 23 h by shortening the waiting time between repetitions and by iterating over the time points first to reduce the effect of accumulating heat. Still, these relatively long experiments were often affected by fluctuations in the environment. One of these data sets is shown in Fig. 2.13b. This data set with 47 time points and 5000 averages was recorded in 41 h (error in Ref. [21]) using a longer waiting time between repetitions than in panel b. In addition, the cooling of the TLSs was measured interleaved. Interestingly, the curves can be grouped into at least four families, suggesting different configurations of the TLS environment. The relaxation dynamics is mainly affected by the most resonant TLSs, which give the main contributions to the cross relaxation and contribute to the relaxation in particular at the beginning of the relaxation curves. The data set shown in the panel b was selected to be one of the few without fluctuations.

Drifts and fluctuations during the relatively long measurement time complicate the accurate description of the data with a theoretical model. In addition, during the polarization and the free decay the qubit experiences different decoherence processes, as the readout is not perfectly QND (s. Sec. 2.9). These limitations can be overcome by a much faster measurement scheme based on stroboscopic quantum jumps, which will be introduced in the next section. Since the data acquisition time for quantum jumps experiments is more than ten times faster, such fluctuations are less likely to be observed. Nevertheless, in some of these quantum jump experiments we observe rapid fluctuations of the non-Poissonian quantum jump distributions in great similarity to Ref. [53]. These fluctuations are presented and discussed in Sec. 3.6.6 and are identified as fluctuations of the single most resonant TLS.

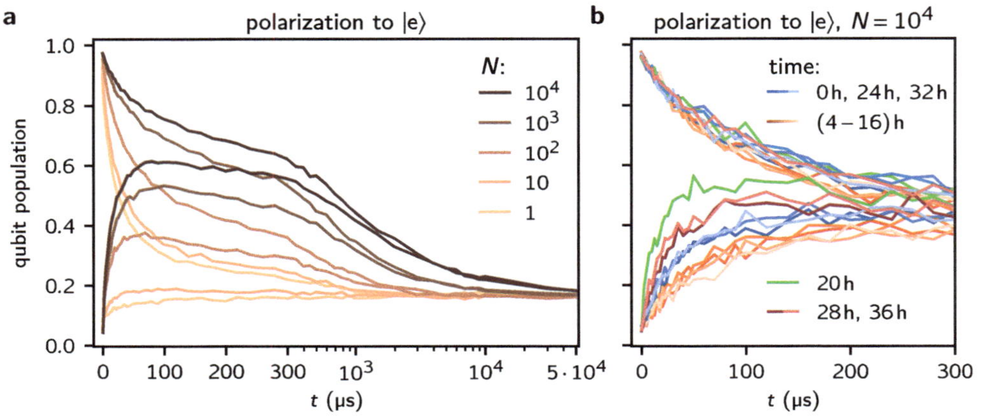

Figure 2.14: Free decay of the qubit after running the Szilard engine polarizing to |e⟩. **a** Free decay of the qubit after polarization to |e⟩ for various times $N \cdot t_{\mathrm{rep}}$ followed by an initialization to either |g⟩ or |e⟩. Note the logarithmic x-axis from 300 μs onwards, which is required to depict the slow relaxation dynamics. The qubit reaches a population inversion of ~ 60 % after being initialized in |g⟩. The curves are the mean of 10 iterations with in total 10000 single shot qubit measurements per data point recorded within 23 h. **b** In some data sets, fluctuations at the beginning of the relaxation curves can be observed. The color coding is used to indicate quantitatively different time evolutions, which can be grouped in at least four families (blue, orange, green and red), suggesting at least four different configurations of the TLS environment. Following Eq. 2.21, the main contributions can be expected from the most resonant TLSs that affect in particular the beginning of the relaxation curves. A corresponding behavior for each trace index is also observed for the curves after polarization to |g⟩, which are measured interleaved (not shown). The data points are based on 5000 averages measured in 41 h and with a longer waiting time between repetitions as in panel b. The data set shown in the left panel was selected to be one of the few without fluctuations. Reproduced from Ref. [21].

2.7.2 Stroboscopic quantum jump measurements

The free decay measurement presented in the previous section has the disadvantages of being relatively slow as every time point has to be recorded separately. To speed up the measurements, we can exploit the fact that the qubit readout is more than 96 % quantum non-demolishing (s. Fig. 2.7c, Sec. 2.9 and Ref. [33]). This means that, on average, the populations of the density matrix are not significantly affected by the readout pulse, and we can perform repeated single shot qubit measurements to monitor the qubit relaxation. The control sequence is depicted in Fig. 2.15a. The repetition time of the qubit measurements during the polarization and qubit monitoring was chosen to be the same, such that the qubit experiences the same decoherence processes. The main advantage of the stroboscopic jump method is the possibility to determine the qubit's transition rates $\Gamma_{\uparrow,\downarrow}$, which allows to distinguish between changes in the energy relaxation rate and changes in the equilibrium population of the qubit.

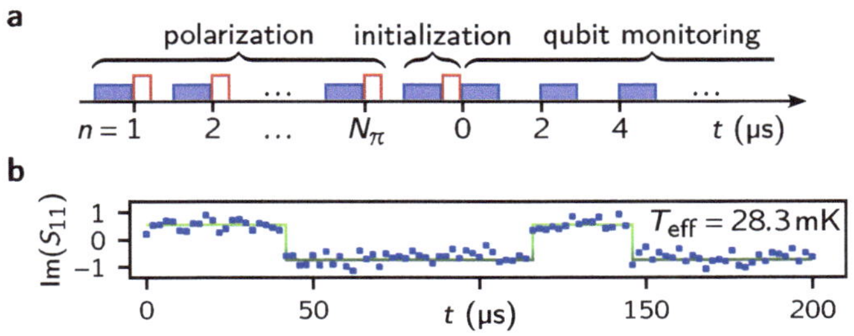

Figure 2.15: Control sequence implementing the Szilard engine and stroboscopic quantum jumps. **a** Control sequence similar to Fig. 2.13 except that the qubit relaxation is measured with stroboscopic quantum jumps. We use the same measurement repetition time of $t_{\text{rep}} = 2\,\mu\text{s}$ for the polarization and the qubit monitoring. **b** Typical stroboscopic quantum jump trace during qubit monitoring (as shown in panel a). The solid line indicates the assigned qubit state. Reproduced from Ref. [21].

A typical quantum jump trace is shown in Fig. 2.15b. Here, the qubit was initially measured in the excited state. After some time, the qubit is measured in the ground state due to qubit relaxation to thermal equilibrium, and from time to time it is measured again in the excited state due to the finite temperature. The average qubit population defines the thermal equilibrium population $p_{\text{th}} = 12\,\%$, corresponding to a temperature of $T_{\text{eff}} = 28.3\,\text{mK}$. This temperature should be compared with the thermal population of the free decay measurements (Fig. 2.14a), where we find $p_{\text{th}} = 16\,\%$ corresponding to the true qubit temperature $T = 34\,\text{mK}$. For comparison, the temperature of the dilution refrigerator is $\sim 25\,\text{mK}$.

Fig. 2.16 and Fig. 2.17 show measured qubit relaxation curves for the various polarization and initialization scenarios. The 20 different scenarios were measured interleaved in 25 iterations with in total 2500 repetitions. The 100 repetitions in each iteration were separated by additional 50 ms for further relaxation of the system. In retrospect, this waiting was not needed. It would probably be better to continue the stroboscopic measurements to ensure a constant effective temperature (Sec. 2.9). The entire data set was recorded within 3.5 h. Approximately half of the time was used to transfer the data of each iteration from the FPGA control server to the measurement computer. The theoretical model plotted on top of the data is discussed in Sec. 2.8.

As already demonstrated for the free decay measurements (Fig. 2.14a), after sufficiently long polarization times to the excited state the qubit reaches population inversion ($N \geq 10^3$, Fig. 2.16b, right). Here, the population inversion is slightly less pronounced, potentially due to the cooling influence of the stroboscopic measurements. However, the population inversion can be confirmed even more directly by the inversion of the transition rates $\Gamma_\uparrow > \Gamma_\downarrow$ (Fig. 2.20a in Sec. 2.7.5). A notable consequence is that for $N = 10^4$ the preparation fidelity for the excited state is higher than for the ground state (Fig. 2.20a, inset). Therefore, measuring the active reset fidelity with a fast repetition time, without letting the system relax to thermal

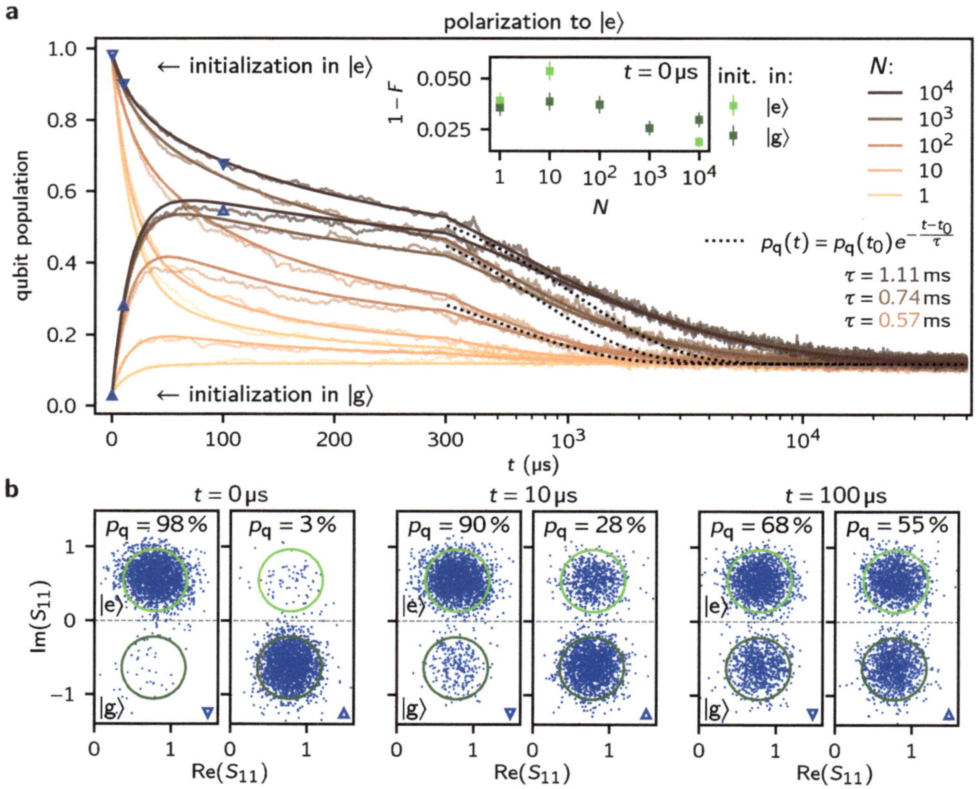

Figure 2.16: Qubit evolution after running the Szilard engine. **a** Measured relaxation of the qubit after polarization to |e⟩ for various times $N \cdot t_{\text{rep}}$ followed by an initialization to either |g⟩ or |e⟩. Note the logarithmic x-axis from 300 μs onwards.The exponential decay curves shown in dotted lines, with the decay times indicated by the corresponding labels, are guides to the eye to illustrate the highly non-exponential relaxation of the environment (Fig. 3.10). The inset shows the preparation infidelity of the initialization. We observe an increasing fidelity with N, in particular for the initialization in |e⟩. The errorbars show the one sigma confidence intervals of the binomial distribution with 2500 repetitions. **b** Scatter plots of the complex reflection coefficient S_{11} for the relaxation curves shown in panel a for $N = 10^4$. The left panels illustrate the reduced relaxation of the excited state population vs. time. The right panels demonstrate that the qubit undergoes a population inversion due to interactions with the environment. Notably, the |f⟩-state is not populated, as illustrated by the absence of a third cloud in the S_{11} distribution (cf. Fig. 2.18c). The continuous lines correspond to the theoretical model of Eqs. 2.22 and 2.23, applied to all measured curves simultaneously. Reproduced from Ref. [21].

equilibrium, can lead to an overestimation of the active reset fidelity. Last, it should be noted that the population inversion does not involve higher qubit states, as illustrated by the absence of the |f⟩-state population. The |f⟩-state becomes only noteworthily populated for experiments at higher temperatures (cf. Fig. 2.18c).

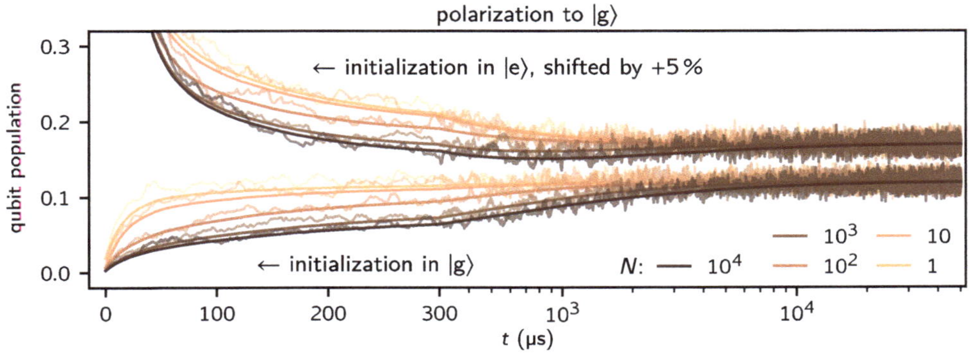

Figure 2.17: Qubit evolution after running the Szilard engine. Measured relaxation of the qubit after polarization to $|g\rangle$ followed by an initialization to either $|g\rangle$ or $|e\rangle$. Compared to Fig. 2.16a, the opposite effect is visible: the environment is cooled by the polarization sequence, demonstrating that the heat flow in the environment is not the trivial result of heating due to repeated microwave readout and control pulses. The upper curves are shifted upwards by 5 % for better visibility. The theoretical curves use the same parameters as that in Fig. 2.16. Reproduced from Ref. [21].

Before continuing, it should be emphasized that the relaxation shows clear signatures of a long-lived TLS environment. On the one hand there is the population inversion, and on the other hand the temperature-independent relaxation time that is shown in Fig. 2.20b. However, it could still be argued that the population inversion is the result of a trivial heating process. For example, the environment could, to a large extend, be heated by the microwave control pulses and the population inversion by a few percent on top could be the result of a demolishing qubit readout.

The bullet-proof experiment that disproves this hypothesis is to cool the environment using same control pulses by polarizing the qubit to the ground state, in which case the qubit becomes the only heat sink in the system. The experimental results showing the cooling of the reservoir are depicted in Fig. 2.17 for various polarization and initialization scenarios. Note that for $N = 10^4$ and after initialization to $|e\rangle$ a non-monotonic relaxation can also be observed. The qubit reaches its minimal population after approximately one millisecond before warming up again.

2.7.3 Hyperpolarization at higher temperature

Due to the relatively low thermal qubit population, the cooling in Fig. 2.17 is less pronounced than the heating of the TLSs in Fig. 2.17a. For this reason, the hyperpolarization was also measured at higher fridge temperature. Fig. 2.18 shows the results for a fridge temperature set to 75 mK. The hyperpolarization is still visible at a fridge temperature of 150 mK (not shown), but much less pronounced, most likely due to a vanishing intrinsic qubit lifetime. In contrast, an increasing

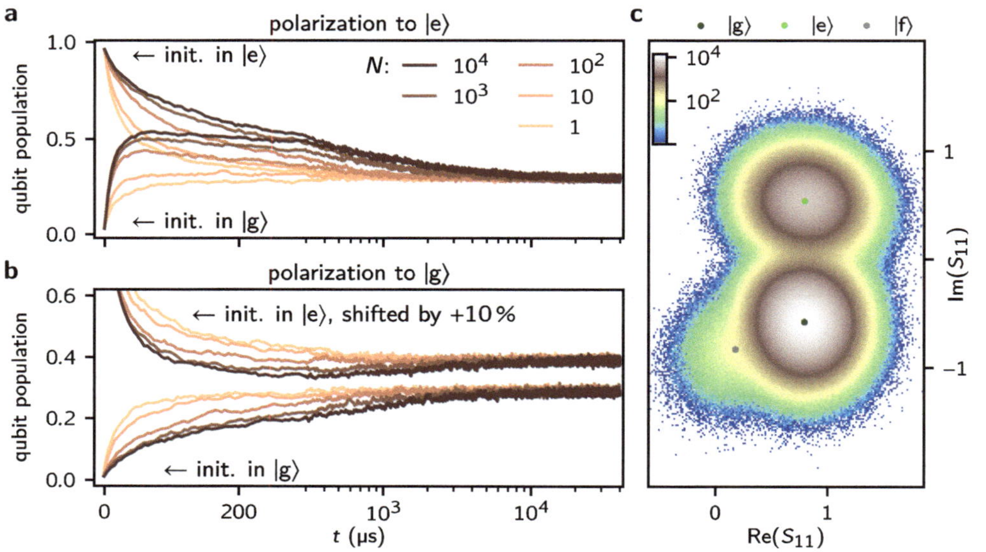

Figure 2.18: Relaxation measurements at T = 75 mK. a, b Qubit relaxation after polarization to $|e\rangle$ and $|g\rangle$, respectively, for various times $N \cdot t_{rep}$ followed by an initialization to $|g\rangle$ or $|e\rangle$. Note that at 75 mK the final thermal population is higher, therefore increasing the visibility of the cooling effect in comparison to Fig. 2.17. The curves corresponding to initialization in $|e\rangle$ are shifted upwards by 10 % for better visibility. For the stroboscopic qubit monitoring the repetition time was set to $t_{rep} = 5\,\mu s$ in order to decrease quantum demolition effects. **c** Histogram of the complex reflection coefficient S_{11} acquired from the last millisecond of the relaxation curves shown in panel a and b. For the thermal populations of the states $|g\rangle$, $|e\rangle$ and $|f\rangle$ we expect (67.7, 31.9, and 0.4) % and we measure (70.6 ± 0.2, 28.8, and 0.4 ± 0.2) %. For $|g\rangle$ and $|e\rangle$ the discrepancy is likely due to quantum demolition effects caused by the readout (s. Sec. 2.9). The populations are obtained from a Gaussian mixture model fit to the S_{11} distribution using the scikit-learn library [80]. Reproduced from Ref. [21].

dephasing of the qubit and the TLSs with temperature should not affect the qubit cross relaxation into the TLS environment (s. Sec. 3.5.3 and Fig. 3.11).

2.7.4 Extraction of the transition rates

The common method for extracting qubit transition rates from quantum jump traces is to analyze the traces sequentially in time. This means that a latching filter is first applied to handle transient or ambiguous measurement points. Then, $\Gamma_{\uparrow,\downarrow}$ can be inferred from the average qubit dwell times in the ground and excited state. The main drawback of this approach is that the time resolution is limited by the time required for a jump to occur.

In contrast, in the procedure presented here, we analyze the sufficiently large number M of quantum jump traces at each time index k in the sequence. This allows

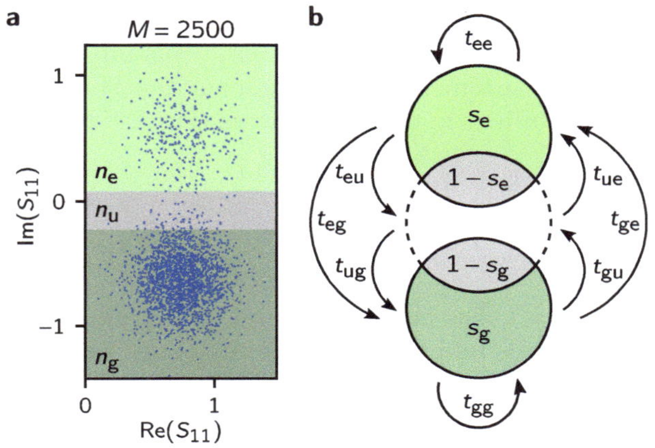

Figure 2.19: Extraction of qubit transition rates. **a** Typical measured S_{11} scatter plot extracted for a given time point from the M stroboscopic quantum jump traces. The three regions discriminate between areas in the S_{11} plane where we declare the qubit state with certainty larger than 99.98 %, highlighted in dark green for $|g\rangle$ and light green for $|e\rangle$, and where the qubit state is subject to state discrimination errors, highlighted in gray. For measurements in the gray area the state is declared to be undetermined. **b** Between two successive sets of M measurements we compute the transition probabilities t_{ij} between the three regions. Assuming the qubit is perfectly polarized in $|g\rangle$ or $|e\rangle$, the probabilities for measurements to be declared with certainty are given by s_g and s_e, respectively. Reproduced from Ref. [21].

for a precise and fast extraction of the time-evolving population and transition rates. However, the method is subject to certain conditions. Most importantly, the state discrimination must be large enough so that most measurements can be unambiguously assigned to a qubit state and the amount of false state assignments is negligible. Next, the probability of missing a quantum jump must be small ($t_{rep} < T_1$). Furthermore, the probability of measuring transient dynamics must be small. Ideally, the measurement is a short (compared to T_1) but strong projective readout pulse. In addition, it will be assumed here that the $|f\rangle$-state population is negligible. If the conditions are met, the method is valid for arbitrary squeezed pointer states and the accuracy is only limited by the available statistics. For finite size environments, however, there is a systematic error in the determination of the transition rates. The stroboscopic qubit measurements lead to correlated fluctuations in the environment, as will be discussed in more detail in Sec. 3.

The error we have to deal with is the state discrimination error due to the finite signal, meaning that the pointer states of the complex reflection coefficient S_{11} overlap (Fig. 2.18c). Three regions in the complex plane are defined corresponding to the qubit in the ground or excited state, or to an undecided state (Fig. 2.19a). The regions are chosen in such a way that the probability to mistake the qubit's ground and excited state can be neglected (in our case it may happen in ~ 0.2 ‰ of the measurements). From the measured $M = 2500$ stroboscopic quantum jump traces, one can obtain a distribution similar to the one shown in Fig. 2.19a at each time stamp k, and with that the three population probabilities p_g, p_e and p_u. By comparing successive measurement results in each of the M quantum jump traces the nine transition probabilities t_{ij} can be extracted (Fig. 2.19b). In other words,

$$\begin{pmatrix} p_g \\ p_e \\ p_u \end{pmatrix}_{k+1} = \begin{pmatrix} t_{gg} & t_{eg} & t_{ug} \\ t_{ge} & t_{ee} & t_{ue} \\ t_{gu} & t_{eu} & t_{uu} \end{pmatrix} \cdot \begin{pmatrix} p_g \\ p_e \\ p_u \end{pmatrix}_{k}, \tag{2.15}$$

with $\sum_i p_i = 1$ and $\sum_i t_{ij} = 1$. The latter, summing the columns of t_{ij}, simply states that the system can either be found in the same region as previously measured or in one of the two other regions. Therefore, out of these three population and nine transition probabilities only eight are independent values. Moreover, given the 5.6 σ state separation in our measurement, the probability to measure state u is small ($1 - s_{g,e} \approx 2.5\,\%$), the rates t_{ug} and t_{ue} have low statistical weight and will not be used in the analysis. Nevertheless, we are left with six values, which are sufficient to extract the five parameters of interest: the qubit population p_q, the probabilities $P_{|g\rangle,|e\rangle}$ and $P_{|e\rangle,|g\rangle}$ that the qubit has jumped up or down, and the probabilities s_g and s_e, which declare the qubit states with certainty. Instead of computing the most likely set of parameters, there is fortunately a much simpler solution. We have

$$\begin{aligned} t_{gg} &= (1 - P_{|g\rangle,|e\rangle}) s_g, & t_{eg} &= P_{|e\rangle,|g\rangle} s_g, \\ t_{ge} &= P_{|g\rangle,|e\rangle} s_e, & t_{ee} &= (1 - P_{|e\rangle,|g\rangle}) s_e, \end{aligned} \tag{2.16}$$

which can be inverted to give

$$s_g = \frac{t_{gg} t_{ee} - t_{ge} t_{eg}}{t_{ee} - t_{ge}}, \qquad s_e = \frac{t_{gg} t_{ee} - t_{ge} t_{eg}}{t_{gg} - t_{eg}}. \tag{2.17}$$

Since the probabilities s_g and s_e are time-independent, they can be averaged along the time trace to achieve a higher precision. Finally, s_g and s_e are used to obtain the successive quantum jump probabilities corrected for state discrimination errors:

$$P_{|g\rangle,|g\rangle} = 1 - P_{|g\rangle,|e\rangle} = t_{gg}/s_g, \qquad P_{|e\rangle,|e\rangle} = 1 - P_{|e\rangle,|g\rangle} = t_{ee}/s_e, \tag{2.18}$$

yielding the transition rates

$$\Gamma_\uparrow = -\log P_{|g\rangle,|g\rangle}/t_{rep}, \qquad \Gamma_\downarrow = -\log P_{|e\rangle,|e\rangle}/t_{rep}. \tag{2.19}$$

For the qubit population one may either use $p_q = 1 - p_g/s_g$ or $p_q = p_e/s_e$. Ideally, one chooses the regions such that $s_g = s_e$, in which case the most likely estimator for the qubit population is

$$p_q = \frac{n_e}{n_g + n_e}. \tag{2.20}$$

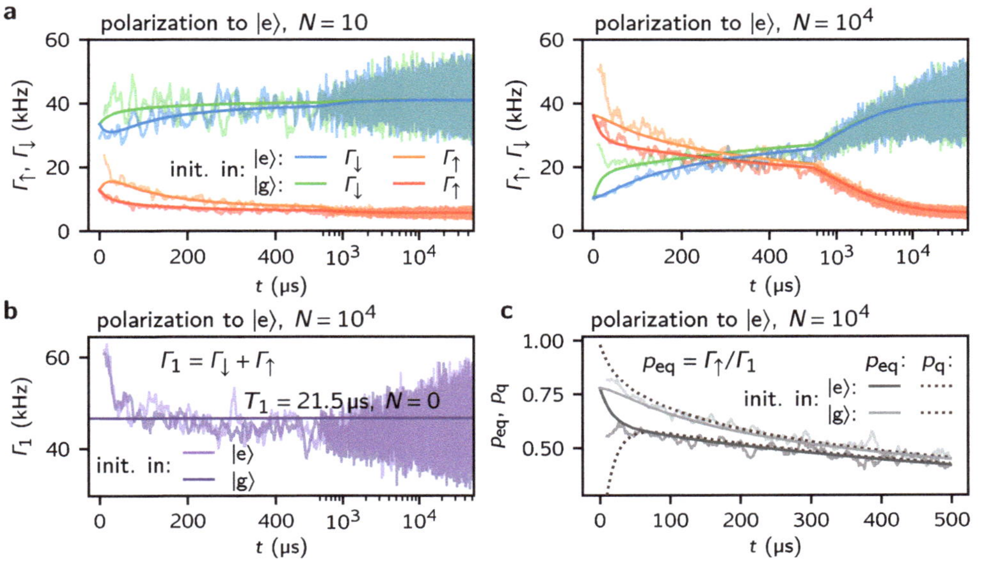

Figure 2.20: Population inversion and constant relaxation rate, signatures of the TLS environment. **a** Measured (light color) and calculated (dark color) qubit transitions rates $\Gamma_{\uparrow,\downarrow}$ following initialization to $|g\rangle$ or $|e\rangle$ and polarization to $|e\rangle$ with $N = 10$, to 10^4 shown in the left and right panel, respectively. The measured rates are extracted from the same quantum jump traces used to extract the qubit relaxation (Fig. 2.16a), with the logarithmic time-axis starting at 500 µs. For long polarization times the rates are reversed in the beginning, meaning that the qubit sees a negative temperature environment. Note that in all cases the $|g\rangle$-state initialization visibly cools the environment, suggesting a heat capacity of only a few energy quanta. To reduce the statistical noise, a five-point moving average filter was applied, corresponding to a 10 µs window. Furthermore, the first 10 µs of the orange and green curves are omitted due to the low statistics; it is unlikely to have two consecutive $\lesssim 5$ µs intervals between jumps. Similarly, these rates are overestimated in the beginning by preferably detecting short T_1 fluctuations of the qubit. **b** Relaxation time Γ_1 obtained from the $\Gamma_{\uparrow,\downarrow}$ rates in panel a for $N = 10^4$. The Γ_1 rate is constant compared to the changes in $\Gamma_\uparrow$ and $\Gamma_\downarrow$ shown in panel a, i.e. Γ_1 is independent of the environmental populations, which indicates a TLS environment. **c** Equilibrium population of the qubit p_{eq} extracted from the $\Gamma_{\uparrow,\downarrow}$ rates. The dashed lines show the corresponding qubit population p_q relaxation taken from Fig. 2.16a. We extrapolate an effective population of the environment $p_{eq} = 78\,\%$ at $t = 0$. In all panels, the theoretical curves use the same parameters as that in Fig. 2.16. Reproduced from Ref. [21].

2.7.5 Negative temperatures and constant relaxation time

As was explained in the previous section, the time-evolving transition rates (Fig. 2.20a) are obtained from the stroboscopic quantum jump traces by using $\Gamma_\uparrow = -\ln\big(P_{|g\rangle,|g\rangle}\big)/t_{rep}$ and $\Gamma_\downarrow = -\ln\big(P_{|e\rangle,|e\rangle}\big)/t_{rep}$, where P is the probability to measure the same qubit state in successive measurements. These rates define the relaxation rate $\Gamma_1 = \Gamma_\uparrow + \Gamma_\downarrow$ and the equilibrium population of the qubit $p_{eq} = \Gamma_\uparrow/\Gamma_1$ (s.

Sec. 2.8). Note that the noise magnitude varies with the qubit population (Fig. 2.20a) because the rates $\Gamma_{\uparrow,\downarrow}$ are based on conditional probabilities.

Remarkably, after a heating sequence with $N = 10^4$, the qubit relaxation rate Γ_1 is comparably constant (Fig. 2.20b); in contrast p_{eq} follows a non-exponential relaxation for time scales up to 50 ms. From the overlayed theoretical model, we can ascribe a hyperpolarization of $p_{\mathrm{eq}}^{\mathrm{TLSs}} = 97\,\%$ for the TLSs at the end of the polarization sequence, which, when taking into account the intrinsic loss of the qubit, gives the measured $p_{\mathrm{eq}} = 78\,\%$ (Fig. 2.20c, and Eqs. 3.70 and 3.71). Conversely, after a cooling sequence with $N = 10^4$, $p_{\mathrm{eq}} = 3.0\,\%$ is extracted, as can be ascertained in Fig. 2.17 using that the qubit population $p_{\mathrm{q}} = 2.0\,\% \approx p_{\mathrm{eq}}$ after $1/\Gamma_1$. Hence, the Szilard engine cooled the environment to an effective temperature of 16 mK, which is well below the ~ 25 mK temperature of the dilution refrigerator and the effective temperature $T_{\mathrm{eff}} = 28.3\,\mathrm{mK}$ corresponding to the qubit population $p_{\mathrm{th}} = 12.0\,\%$ (s. Sec. 2.7.2 and Fig. 2.23). The TLS hyperpolarization is even lower, $p_{\mathrm{eq}}^{\mathrm{TLSs}} = 3.4\,‰ \mathrel{\hat{=}} 9.9\,\mathrm{mK}$, limited by the qubit preparation infidelity. The values are extrapolated from the theoretical model, which will be explained in the next section. For both, heating and cooling, the hyperpolarization values are among the highest reported in literature [81, 82].

2.8 Theoretical modeling

The Solomon equations that describe the relaxation of the qubit coupled to an arbitrary number of TLSs is derived in full rigor in Chpt. 3. The main results thereof are used here to model and interpret the experimental findings.

The constant relaxation rate Γ_1 as well as the observed population inversion indicate an environment consisting of TLSs. The system is therefore modeled assuming the qubit to be coupled to a countable number of TLSs with populations p_{t}^k. The cross relaxation rates Γ_{qt}^k between the qubit and the TLSs are given by [83, 84]

$$\Gamma_{\mathrm{qt}}^k = \frac{2g^2\Gamma_2}{\Gamma_2^2 + \delta_k^2}, \tag{2.21}$$

where δ_k is the detuning between the qubit and the kth TLS, g their transverse coupling strength, and Γ_2 the sum of their decoherence rates. Since the TLSs can in turn excite the qubit, we conclude that the qubit and the TLSs are close in frequency. This means that they relax approximately to the same thermal population p_{th} (note that the qubit is well thermalized, as discussed earlier). Finally, we introduce intrinsic relaxation rates for the qubit and the TLSs, Γ_{q} and Γ_{t}^k, respectively, capturing the remaining environment (Fig. 2.12a).

The relaxation dynamics is governed by the so-called Solomon equations [83], extensively used in the field of nuclear hyperpolarization [85]. The equations read

$$\dot{p}_\mathrm{q} = -\Gamma_\mathrm{q}(p_\mathrm{q} - p_\mathrm{th}) - \sum_k \Gamma^k_\mathrm{qt}(p_\mathrm{q} - p^k_\mathrm{t}) \tag{2.22}$$

$$\dot{p}^k_\mathrm{t} = -\Gamma^k_\mathrm{t}(p^k_\mathrm{t} - p_\mathrm{th}) - \Gamma^k_\mathrm{qt}(p^k_\mathrm{t} - p_\mathrm{q}), \tag{2.23}$$

where we identify the constant qubit relaxation rate $\Gamma_1 = \Gamma_\mathrm{q} + \sum_k \Gamma^k_\mathrm{qt}$ and the time-dependent equilibrium population

$$p_\mathrm{eq} = \left(\Gamma_\mathrm{q} p_\mathrm{th} + \sum_k \Gamma^k_\mathrm{qt} p^k_\mathrm{t}\right) / \Gamma_1. \tag{2.24}$$

During this polarization sequence of length $N \cdot t_\mathrm{rep}$, the qubit population is approximately constant, corresponding to the targeted preparation state. As a consequence of Eq. 2.23, the Solomon equations predict an exponential relaxation for the TLSs to their new steady-state value. Immediately after the polarization sequence at $t = 0$, the TLS populations $p^k_{\mathrm{t},0} := p^k_\mathrm{t}(t = 0)$ are given by

$$p^k_{\mathrm{t},0} = \left(p_\mathrm{th} - p^k_\mathrm{s}\right) e^{-\left(\Gamma^k_\mathrm{qt} + \Gamma_\mathrm{t}\right) N t_\mathrm{rep}} + p^k_\mathrm{s} \tag{2.25}$$

with steady-state values

$$p^k_\mathrm{s} = \frac{\Gamma_\mathrm{t} p_\mathrm{th} + \Gamma^k_\mathrm{qt}}{\Gamma_\mathrm{t} + \Gamma^k_\mathrm{qt}} \quad \text{or} \quad p^k_\mathrm{s} = \frac{\Gamma_\mathrm{t} p_\mathrm{th}}{\Gamma_\mathrm{t} + \Gamma^k_\mathrm{qt}} \tag{2.26}$$

for heating or cooling of the environment, respectively. These initial TLS populations $p^k_{\mathrm{t},0}$ are shown in Fig. 2.21 for various polarization times, assuming that the TLSs are equally spaced in frequency with otherwise identical parameters. In view of this initial hyperpolarization the polarization sequence may also be interpreted as a spectral hole burning. The extraction of the system parameters that lead to this distributions are discussed in the next paragraph.

So far, the model in Eqs. 2.22 and 2.23 requires two rates for each TLS. In order to extract meaningful information from the measurements by virtue of Eq. 2.21 simplifying assumptions have to be made and the number of fitting parameters needs to be reduced. Since the TLS polarization is observed in different qubits and at different qubit frequencies (s. Supp. Ref. [21]) the TLSs are expected to be randomly distributed in frequency. This distribution is simplified by modeling them to be equally spaced in frequency with $\delta_k = k\Delta + \Delta_0$, where $\Delta_0 \in [0, \Delta/2]$ defines a shift of the TLS ladder with respect to the qubit frequency. This is justified by the fact that we are mainly interested in capturing the slow, non-exponential relaxation at millisecond timescales. With the same argument, we can assume the same g and

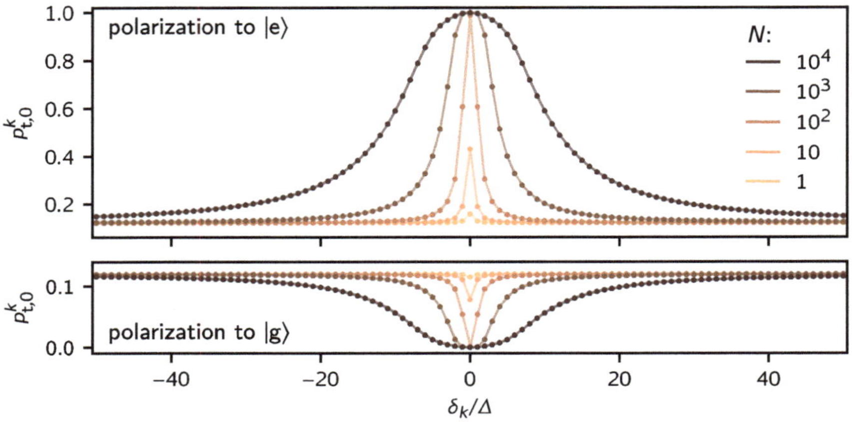

Figure 2.21: TLS hyperpolarization. Initial TLS populations $p^k_{t,0}$ plotted versus their detuning δ_k for various polarization times $N \cdot t_{rep}$, with t_{rep} being the repetition time of the stroboscopic qubit preparations to $|e\rangle$ (top panel) and $|g\rangle$ (bottom panel). The TLS populations were calculated using Eq. 2.25. Reproduced from Ref. [86].

Γ_2 for all TLSs. The price being payed using these simplification is that the model captures less accurately the initial features of the decay curves, at $t < 300\,\mu s$. Indeed, these features are a fingerprint of the exact TLS configuration, and they fluctuate in time, as can be seen in Fig. 2.14b.

The simplified model allows to rewrite Eq. 2.21 in the form $\Gamma^k_{qt} = ab^2/[b^2 + (k + bc)^2]$, showing that g, Δ, Δ_0 and Γ_2 do not appear independently in the model. Instead, $g = \sqrt{a\Gamma_2/2}$, $\Delta = \Gamma_2/b$ and $\Delta_0 = c\Gamma_2$ can be determined for a given decoherence rate from a successful fit of the model. The fit procedure is further restricted by inserting the measured qubit relaxation rate $\Gamma_1 = 1/21.5\,\mu s$ (Fig. 2.20b), leaving us with only two essential fit parameters Γ_q and b. The parameter c has only a minor influence at the beginning of the relaxation curves and leads to a rescaling of the other parameters to keep the qubit relaxation constant (s. Supp. in Ref. [21]). As will be discussed in more detail in Chpt. 3.6.5, for a finite size heat reservoir there is a systematic error when determining the qubit transition rates from quantum jumps. In Chpt. 3 the rate $\Gamma_1 = 1/18.5\,\mu s$ is used, which agrees a bit better with the measured non-Poissonian quantum jump statistics and describes equally well the measured relaxation curves.

The robustness of the model is illustrated by the fact that a fit of only the first millisecond to one of the stronger polarized relaxation curves (e.g. polarization to $|e\rangle$ for $N = 10^3$ with initialization in $|g\rangle$ or $|e\rangle$) is sufficient to describe the highly non-exponential relaxation of all measurements on the entire relaxation range up to 50 ms (Fig. 2.16a, Fig. 2.17, Fig. 2.20 and Fig. 3.10c in Chpt. 3). Further details of the fitting procedure are presented in the Supplementary of Ref. [21].

Using the lower bound $\Gamma_2 \geq \Gamma_2^{\mathrm{q}} \approx 0.5\,\mathrm{MHz}$, where Γ_2^{q} is the decoherence rate of the qubit (Sec. 2.3.2), $g \geq 2\pi \cdot 12\,\mathrm{kHz}$ and $\Delta \geq 2\pi \cdot 167\,\mathrm{kHz}$ are extracted. In Chpt. 3, a TLS decoherence of $\Gamma_2^{\mathrm{t}} = 1.0\,\mathrm{MHz}$ is used for most of the quantum jump simulations. In this case, we have $g = 2\pi \cdot 22\,\mathrm{kHz}$ and $\Delta = 2\pi \cdot 500\,\mathrm{kHz}$. The comparably small coupling strength $g \ll \Gamma_2^{\mathrm{q}}$ is consistent with the fact that no avoided level crossings in the qubit spectrum are observed. Notably, this argument remains valid even for higher decoherence because g and Δ scale with $\sqrt{\Gamma_2}$ and Γ_2, respectively. Using an upper bound for the decoherence $\Gamma_2 \sim 1/10\,\mathrm{ns} \ll f_{\mathrm{q}}$, comparable to values reported in Ref. [65] for dielectric TLSs, gives $g < 2\pi \cdot 170\,\mathrm{kHz}$ and $\Delta < 2\pi \cdot 35\,\mathrm{MHz}$.

Furthermore, the two contributions of the qubit relaxation can be calculated: one rate is due to interactions with the TLSs, $\Gamma_{\mathrm{qt}}^{\mathrm{TLSs}} = \sum_k \Gamma_{\mathrm{qt}}^k = 35.9\,\mathrm{kHz}$, and the other is the remaining intrinsic relaxation $\Gamma_{\mathrm{q}} = 10.7\,\mathrm{kHz}$. The TLS bath is therefore identified as the dominant loss mechanism. Remarkably, the fit also indicates that the intrinsic relaxation time exceeds $1/\Gamma_{\mathrm{t}} \geq 50\,\mathrm{ms}$, which is orders of magnitude longer than previously measured relaxation rates of dielectric TLSs [64, 66, 87]. This fact leads us to believe that we are reporting a new type of TLS environment, possibly related to spins [52, 88] or trapped quasiparticle TLSs [89]. Finally, we would like to mention that $\Gamma_{\mathrm{qt}}^k \geq \Gamma_{\mathrm{t}}$ for $|k| \leq 15$, which means that the qubit is the main decay channel for at least the first few tens most resonant TLSs.

Following Szilárd's seminal paper [22], the homonymous engine uses measured information as fuel (Sec. 2.2). In the first iteration of a cooling sequence starting from thermal equilibrium $T = 28.3\,\mathrm{mK}$, the engine extracts on average the internal energy $\Delta U = 0.24\,k_{\mathrm{B}}T$ from the qubit, corresponding to an entropy reduction of $\Delta S_{\mathrm{q}} = 0.37\,k_{\mathrm{B}}$, which should be compared with the entropy produced by the measurement apparatus $k_{\mathrm{B}} \ln 2 \approx 0.69\,k_{\mathrm{B}}$ (Sec. 2.2). From the rate equation we can calculate the optimal working regime for our Szilard engine. Using the fitted parameters we infer that the maximum heat reduction $\Delta Q_{\mathrm{R}} = 0.11\,k_{\mathrm{B}}T$ in the reservoir occurs 68 µs after the qubit initialization. Thus, at most half of the extracted heat from the qubit can be used to cool the reservoir. This surprisingly small value seems to be in conflict with the TLS bath being the dominant loss mechanism $\Gamma_{\mathrm{qt}}^{\mathrm{TLSs}} \approx 3.3\Gamma_{\mathrm{q}}$. This discrepancy can simply be explained by the finite size of the reservoir. The qubit only interacts strongly with the few most resonant TLSs. Consequently, when the qubit is reset to its ground state, the temperature of these TLSs will reduce and the qubit can not reach its prior energy, thus $\Delta Q_{\mathrm{R}} < \Delta U_{\mathrm{q}}(T_{\mathrm{R}})$. The finite size and the effect of a single π-pulse is best visible at higher and negative temperatures, showing a significant cooling of the TLSs (Fig. 2.20).

2.9 SPAM errors and quantum demolition effects

The following section provides further insight into the polarization process. The analysis can be used to disentangle state preparation and measurement (SPAM)

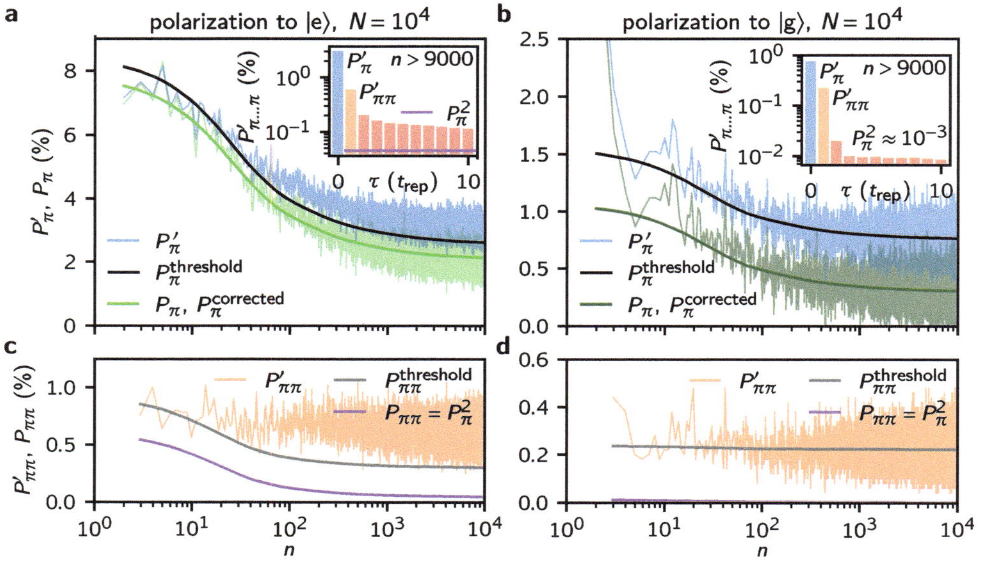

Figure 2.22: Active feedback qubit preparations during the polarization sequence. Measured probabilities P'_π (blue) and $P'_{\pi\pi}$ (yellow) during the polarization sequence to play a π-pulse or respective two successive π-pulses for polarization to $|e\rangle$ (left panels) and polarization to $|g\rangle$ (right panels). The continuous lines in green and violet show the theoretically expected behavior due to TLS polarization. The theoretical curves in black and gray include state discrimination errors. In light green $P_\pi^{\text{corrected}}$ is shown, i.e. the measured P'_π corrected for SPAM errors. The insets show the measured auto-correlation function $P'_{\pi...\pi}(\tau)$ of the polarization sequence averaged over all M experimental repetitions, over $n > 9000$, where the TLS polarization is approximately constant. Reproduced from Ref. [21].

errors from relaxation and may render useful to identify spurious correlations in the feedback control, which can hint at memory effects in the qubit environment.

During the polarization sequence one can observe a decrease in active feedback preparations of the qubit. We quantify the probability to play a π-pulse on the qubit by P'_π, with prime denoting the measured probabilities containing all sorts of errors and correlations. These probabilities are shown in Fig. 2.22 (light blue). In the following we show how to extract the portion P_π (without prime) which originates solely from qubit relaxation.

For comparison, we will first derive the results for an ideal quantum Szilard engine with perfect reset capabilities. In addition, we assume that the polarization of the TLSs follows Eq. 2.25, which in turn gives $p_{\text{eq}}(t)$. We thereby neglect the relaxation of the system in between the measurements. The evolution of P_π during the polarization sequence is explained by the TLS environment that becomes increasingly more polarized. The probability to measure the qubit relaxed from its target state during the repetition time t_{rep} is given by

$$P_\pi(t) = \begin{cases} [1 - p_{\text{eq}}(t)] - [1 - p_{\text{eq}}(t)]e^{-\Gamma_1 t_{\text{rep}}}, & \text{for polarization to } |\text{e}\rangle, \\ [p_{\text{eq}}(t) - 0] - [p_{\text{eq}}(t) - 0]e^{-\Gamma_1 t_{\text{rep}}}, & \text{for polarization to } |\text{g}\rangle. \end{cases} \quad (2.27)$$

Since Γ_1 is constant, the polarization of the TLSs is entirely encoded in $p_{\text{eq}}(t)$. Note that the first term in each equation gives the probability to require a π-pulse after the qubit has reached the equilibrium population. The calculated P_π curves are shown in Fig. 2.22 (solid green lines).

For completeness, using $\Gamma_1 = \Gamma_\uparrow + \Gamma_\downarrow$ and $p_{\text{eq}} = \Gamma_\uparrow/\Gamma_1$, Eq. 2.27 can be approximated to reveal the following relation with the transition rates:

$$P_\pi(t) \approx \begin{cases} 1 - e^{-\Gamma_\downarrow(t) t_{\text{rep}}}, & \text{for polarization to } |\text{e}\rangle, \\ 1 - e^{-\Gamma_\uparrow(t) t_{\text{rep}}}, & \text{for polarization to } |\text{g}\rangle, \end{cases} \quad (2.28)$$

with the approximation being of order $O\left(\Gamma_\downarrow \Gamma_\uparrow t_{\text{rep}}^2/2\right)$, corresponding to the probability that a double quantum jump was undetected within t_{rep}. We thus obtain $\Gamma_{\uparrow,\downarrow} = -\log(1 - P_\pi)/t_{\text{rep}}$, which is the formula used in Sec. 2.7.4. We want to remark the following: (i) when Γ_1 is constant and known, one can directly solve Eq. 2.27 and obtain both $\Gamma_{\uparrow,\downarrow}$-rates, (ii) when Γ_1 is unknown one can only extract one of the transition rates for each polarization state by solving Eq. 2.28, and (iii) linearizing both equations gives $\Gamma_{\uparrow,\downarrow} = P_\pi/t_{\text{rep}}$, however, this approach entails an additional quadratic error compared to solving Eq. 2.28.

As mentioned in the introductory paragraph, in the experiment P_π is altered by various errors to give the measured P'_π. In the insets of Fig. 2.22 we show the auto-correlation function $P'_{\pi\ldots\pi}(\tau) = E[P'_\pi(t+\tau)P'_\pi(t)]$ of the π-pulse sequence with the expectation value E taken over all M repetitions, over the last 1000 pulses of the sequence where the TLS polarization reached its steady state. For a well behaved Born-Markovian environment and in the absence of SPAM errors, the π-pulses are uncorrelated. This means that the probability to have two π-pulses in succession equals $P'_{\pi\pi} := P'_{\pi\ldots\pi}(t_{\text{rep}}) = P'^2_\pi$, and more generally $P'_{\pi\ldots\pi}(\tau > 0) = P'^2_\pi$. Instead, as we show in Fig. 2.22, we observe a strong excess probability, in particular for $P'_{\pi\pi}$. Indeed, if the qubit state is falsely detected or if the reset is unprecise, there will be an increased probability to reset the qubit in the next round.

Next, two approaches are presented to explain and correct the SPAM errors. First, we show to which extend the SPAM errors are caused by state discrimination errors, which can easily be included into the model by forward propagation. In the experiment the threshold to play a π-pulse was chosen to be exactly in-between the qubit's pointer states. Therefore, we have a state discrimination error of $p^{\text{error}}_{|\text{g}\rangle} =$ 2.42 ‰ and $p^{\text{error}}_{|\text{e}\rangle} = 2.68$ ‰, which was extracted from a Gaussian mixture model fit to the complex scatter parameter S_{11}. The difference results from the slightly squeezed noise, as can be seen in Fig. 2.14c. For details on the squeezing see Ref. [90]. When these errors are included into model Eq. 2.27, the state discrimination error perfectly explains the polarization to $|g\rangle$ data. However, for polarization to $|e\rangle$ a good agreement can only be observed in the beginning (Fig. 2.22, black and gray curves). At this level, the following effects could become relevant and explain the remaining difference: The qubit TLS relaxation in between the measurements must be considered, in particular after a false detection and reset to the ground state, the latency between measurement and π-pulse leads to additional SPAM errors, and ultimately the measurement back action on the TLSs leads to additional correlations (Sec. 3.6.3).

Since the state discrimination can not account for all SPAM errors and since all the other mentioned errors are difficult to predict, the second approach is to correct the measured P'_π by the measured excess probability of $P'_{\pi\pi}$. Let P_π be the probability that the qubit has to be reset due to its relaxation and let q be the probability that the active feedback step yields a SPAM error. We will assume that an error will be corrected with certainty in the next round, thereby truncating higher order error propagation. Higher order processes where the qubit relaxed and got falsely measured in its target state are also excluded. Under these assumptions the probability to find a π-pulse due to a SPAM error is $P_{\mathrm{SPAM}} = (1 - P_{\mathrm{SPAM}})q$, and hence $P_{\mathrm{SPAM}} = q/(1+q)$, which is also the probability to find a π-pulse that corrects the previous SPAM error. Similarly, the probability to find a π-pulse due to relaxation is $(1 - P_{\mathrm{SPAM}})P_\pi = P_\pi/(1+q)$. Adding these three probabilities gives P'_π, while summing the probabilities of all five combinations resulting in two successive π-pulses gives $P'_{\pi\pi}$. We have

$$P'_\pi = \frac{P_\pi + 2q}{1+q} \tag{2.29}$$

$$P'_{\pi\pi} = \frac{P_\pi^2 + 2qP_\pi + q(1+q)}{1+q}. \tag{2.30}$$

Longer ranged correlations are approximately and increasingly more uncorrelated $P'_{\pi\ldots\pi}(\tau > t_{\mathrm{rep}}) \approx {P'_\pi}^2$. The above equations can be inverted, allowing to disentangle relaxation and SPAM error contributions:

$$P_\pi^{\mathrm{corrected}} := P_\pi = \frac{P'_\pi - (2 - P'_\pi)P'_{\pi\pi}}{(1 - P'_\pi)^2} \tag{2.31}$$

$$q = \frac{P'_{\pi\pi} - {P'_\pi}^2}{(1 - P'_\pi)^2}. \tag{2.32}$$

The corrected $P_\pi^{\mathrm{corrected}}$ curves are shown in Fig. 2.22 and match well the theoretical curves P_π. At the end of the polarization sequence (from the insets) we extract for the SPAM errors $q = 5.5$ ‰ and $q = 2.2$ ‰ for polarization in $|e\rangle$ and $|g\rangle$, respectively. The accuracy of $P_\pi^{\mathrm{corrected}}$ and q is limited by all neglected higher order errors and moreover by all the other unexplained excess probabilities starting with $P_{\pi\ldots\pi}(2t_{\mathrm{rep}})$.

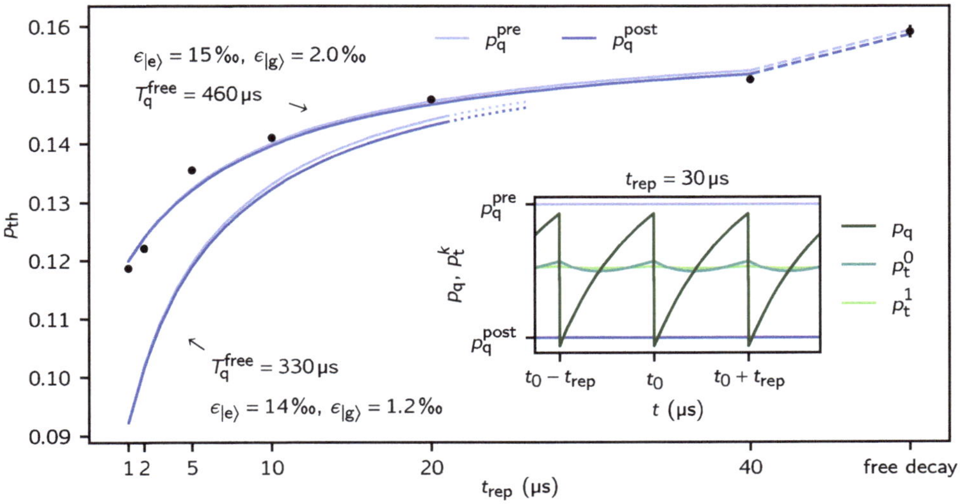

Figure 2.23: Quantum demolition effects induced by the readout. Thermal qubit population p_{th} as a function of the repetition time. The demolition errors $\epsilon_{|e\rangle,|g\rangle}$ of the qubit measurement have a cooling effect on the qubit. When not visible the error bars are in the range of the measured black points. The inset illustrates the periodic steady-state dynamics of the qubit and the TLSs calculated with the Solomon equations. The qubit and TLS populations are enclosed approximately by the qubit populations p_q^{pre} and p_q^{post} before and after the qubit measurement, respectively. The calculated curves modeling the data are based on the assumption that the population of the intrinsic qubit environment, given by the free decay p_{th}, remains unchanged with decreasing repetition time t_{rep}. If the intrinsic qubit environment increases in temperature with the measurement repetition, the lower set of curves could describe the data as well. For more details see text. Reproduced and extended from Ref. [21].

Estimating readout demolishing effects

It should be noted that the intrinsic qubit relaxation Γ_q includes already relaxation processes that originate from demolishing effects of the qubit measurement. The lower true intrinsic qubit relaxation Γ_q^{free} during free decay is visible in the prolonged relaxation of the free decay measurements (Fig. 2.14a). In the Supplementary of Ref. [21], the qubit transition rates for various readout repetition times are contrasted. However, the finite size TLS environment precludes a precise estimation of the qubit transition rates due to the non-Poissonian quantum jump statistics, i.e. correlated fluctuations in the TLS reservoir (Sec. 3.6). The overestimation of the measurement induced relaxation rate in Ref. [21] greatly exceeds the extracted intrinsic qubit relaxation.

In the following, the analysis of Ref. [21] is therefore extended to yield a more precise estimate of the true intrinsic relaxation Γ_q^{free} and the demolition errors $\epsilon_{|e\rangle,|g\rangle}$ from the readout. The information is extracted from the change in thermal population with decreasing readout repetition time. Certainly, interleaved hyperpolarization experiments with different readout repetition times would be even more informative.

Let $p_\mathrm{q}^\mathrm{pre}$ and $p_\mathrm{q}^\mathrm{post}$ denote the qubit population before and after the measurement, respectively. These populations are related by

$$p_\mathrm{q}^\mathrm{post} = (1 - \epsilon_{|\mathrm{e}\rangle})p_\mathrm{q}^\mathrm{pre} + \epsilon_{|\mathrm{g}\rangle}(1 - p_\mathrm{q}^\mathrm{pre}). \tag{2.33}$$

If the repetition time t_rep of the measurements is short compared to all relaxation rates in the system, the relaxation in between the measurements is described by the qubit's true intrinsic lifetime in the steady state. We thus obtain another relation

$$p_\mathrm{q}^\mathrm{pre} = \left(p_\mathrm{q}^\mathrm{post} - p_\mathrm{th}^\mathrm{free}\right) e^{-\Gamma_\mathrm{q}^\mathrm{free} t_\mathrm{rep}} + p_\mathrm{th}^\mathrm{free}. \tag{2.34}$$

We make use here of the fact that the qubit is the dominant loss mechanism for the TLSs, hence the TLS populations will settle to the average qubit population, i.e. $p_\mathrm{t}^k \approx \left(p_\mathrm{q}^\mathrm{pre} + p_\mathrm{q}^\mathrm{post}\right)/2$. It is only when $t_\mathrm{rep} > 1/\max(\Gamma_\mathrm{qt}^k)$ that the non-exponential relaxation begins to make a minor difference. The steady-state dynamics is depicted in the inset of Fig. 2.23. In the free decay limit, when $t_\mathrm{rep} \longrightarrow \infty$, the two equations are again exact with $p_\mathrm{q}^\mathrm{pre}$ approaching the thermal population $p_\mathrm{th}^\mathrm{free}$, which is defined by the temperature of the intrinsic qubit environment. In this limit, the TLS populations also approach $p_\mathrm{th}^\mathrm{free}$.

Solving both equations yields the thermal qubit population as a function of the repetition time. In addition, we can use the constraint $\Gamma_\mathrm{q} = \Gamma_\mathrm{q}^\mathrm{free} + \Gamma_\mathrm{q}^\mathrm{m}$ with the measurement induced qubit relaxation rate $\Gamma_\mathrm{q}^\mathrm{m} = -\log\left(1 - \epsilon_{|\mathrm{e}\rangle}\right)/t_\mathrm{rep} - \log\left(1 - \epsilon_{|\mathrm{g}\rangle}\right)/t_\mathrm{rep}$ and $t_\mathrm{rep} = 2\,\mu\mathrm{s}$, as used in the experiment. From the measured thermal population (Fig. 2.23), we extract the demolition errors $\epsilon_{|\mathrm{e}\rangle} = 15\,‰$ and $\epsilon_{|\mathrm{g}\rangle} = 2.0\,‰$ and the true intrinsic qubit lifetime $T_\mathrm{q}^\mathrm{free} = 460\,\mu\mathrm{s}$, which would be measured in free decay and in the absence of the TLS environment. This lifetime should be compared to the measured intrinsic lifetime $T_\mathrm{q} = 90\,\mu\mathrm{s}$, which shows that the stroboscopic measurements contribute dominantly to the intrinsic qubit loss.

The presented analysis is based on the assumption that the stroboscopic measurements do not heat up the intrinsic qubit environment. If the environment heats up, a lower intrinsic qubit lifetime could also explain the data. In Fig. 2.23, an additional set of curves is shown assuming lower demolishing errors and an increased intrinsic qubit relaxation. In order to explain the measured thermal population, the heating of the intrinsic environment must reach a 25 % population for $t_\mathrm{rep} = 2\,\mu\mathrm{s}$, which corresponds to a temperature of $\sim 50\,\mathrm{mK}$. Since the heating in general and higher temperatures are increasingly unlikely, we conclude that qubit lifetimes of several hundred microseconds can be expected if the TLS environment can be removed in future.

3 Solomon equations

The experiments presented in the previous chapter have revealed the presence of a long-lived TLS environment. The relaxation dynamics of the qubit and the TLSs have been modeled with the Solomon equations. This chapter provides further experimental and theoretical insights into their dynamics. The survey begins with a rigorous derivation of the Solomon equations and ends with the stochastic Schrödinger equation for modeling the non-Poissonian quantum jump statistics of the qubit. The first sections are intended to simplify the introduction to the vast field of open quantum systems and to lay the foundations for the derivation of the Solomon equations. The treatise presented here is an extended and restructured version of Spiecker et. al., Phys. Rev. A 109, 052218, 2024 (Ref. [86]).

3.1 Overview

Relaxation processes induced by spin environments are encountered in many areas of experimental physics. In theoretical physics, they are therefore a popular toy model for understanding decoherence processes and the crossover from the quantum to the classical world [72, 91–103]. A thorough understanding of spin relaxation was first achieved by studying long-lived nuclear spins in the field of nuclear magnetic resonance (NMR). One of the most important NMR spectroscopy methods for the structural analysis of molecules and even proteins [85, 104, 105] is based on the nuclear Overhauser effect and its description via the Solomon equations [83].

The Solomon equations were originally derived for only two interacting nuclear spins [83]. Since the complexity of the relaxation grows exponentially with the number of nuclear spins, the Solomon equations can not be generalized without further ado. In case that the emergence of multispin phenomena contributes significantly to the relaxation process, one may either design the experiment in such a way to suppress their effects or include the dominant cross-correlations between the nuclear spins into the dynamics [106]. Often, these cross-correlations lead only to minor contributions, so that the Solomon equations remain the method of choice to describe the relaxation [85, 104].

Coming from outside the field of NMR, there are several reasons why it is difficult to translate the effect of multispin phenomena to other systems such as our system, i.e. the fluxonium qubit coupled to a large number of long-lived TLSs. First and foremost, one has to overcome the NMR language hurdle: The rates are expressed in

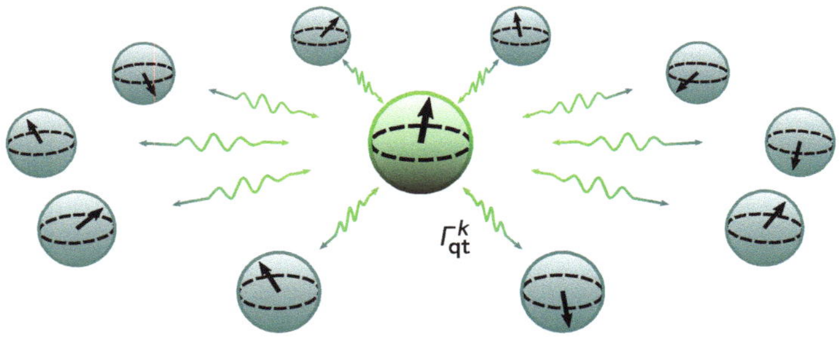

Figure 3.1: Qubit and long-lived TLSs. The qubit acts as the central spin for the surrounding TLSs. Under the presence of decoherence, their interaction leads to a mutual cross-relaxation with individual cross-relaxation rates Γ^k_{qt}.

terms of the correlation time, internuclear distances and orientations in space. The situation is additionally complex due to the vectorial character of the spins, which gives rise to multiple interaction terms, and finally the spins may have even larger quantum numbers, giving rise to quadrupolar spin lattice relaxation [85, 106, 107].

In contrast to NMR, the superconducting qubits and the TLSs are of a very different physical nature. Therefore, we can expect the qubit to interact with a large number of TLSs with potentially different frequencies, coupling strengths and coherence times, as illustrated in Fig. 3.1. For this reason, a new derivation of the Solomon equations is presented for an XX-type interaction and an arbitrary number of TLSs. The derivation starts from a general Lindblad equation [108, 109] for the qubit and the TLSs. To our knowledge, such a derivation of the Solomon equations has not been given in the literature. It turns out that the Solomon equations provide an excellent approximation, in particular for the experimental situation of long-lived TLSs.

In view of the experimental findings presented in Chpt. 2, we show how to deal with Solomon equations of infinite size for a given cross-relaxation distribution and deduce that the relaxation follows a power-law on long time scales. Beyond superconducting devices, the analysis may also prove useful for the accurate description of dipolar relaxation processes, similar to Ref. [110]. Moreover, we contrast the Solomon equations with the widely used Bloch-Redfield master equation, which is applicable when the TLSs form a Born-Markovian environment.

Certainly the main difference to NMR experiments is our ability to measure a single element, the qubit. Instead of looking at the average dynamics, the real-time dynamics can be investigated. This raises the question to what extent it is possible to measure entanglement effects, i.e. cross-correlations between the qubit and the TLSs. As a first step in this direction, we will look at the statistics of the qubit dwell times in its ground and excited state, which is essentially extracted from a single quantum jump trace. We show that the measured non-Poissonian quantum jump statistics of our superconducting qubit can be reproduced by a diffusive stochastic Schrödinger equation. Surprisingly, the quantum jump statistics can also be generated in a much simpler way using the Solomon equations by essentially neglecting the measurement back action on the TLSs. This procedure can therefore serve as a testbed to study the quantum-to-classical transition [14, 111]. The quantum jump analysis can readily be utilized on quantum processors with the TLSs replaced by qubits. A reduced measurement back action could indicate the transition to a chaotic regime as a result of an interaction with a larger Hilbert space, as discussed in Ref. [112].

The system that will be discussed in the following sections consists of a mesoscopic superconducting qubit that interacts with n TLSs via an XX-type coupling, while the TLSs are assumed to be non-interacting. The system Hamiltonian is then given by

$$H = \frac{\hbar\omega_\mathrm{q}}{2}\sigma_\mathrm{q}^z + \sum_{k=1}^{n} \frac{\hbar\omega_k}{2}\sigma_k^z + \sum_{k=1}^{n} \hbar g_k \sigma_\mathrm{q}^x \sigma_k^x, \tag{3.1}$$

where ω_q and ω_k are the frequencies of the qubit and the kth TLS, g_k is their coupling strength and $\sigma_\mathrm{q}^{x,z}$ and $\sigma_k^{x,z}$ denote their Pauli matrices. The physics of the central spin model [113–115] as of Eq. 3.1 with added dissipation is generally complex and an active field of research, including phenomena such as superradiance [116, 117], state revivals [118–120], and exceptional points [121]. With increasing decoherence in the system one can expect a transition from coherent oscillations to a regime where the expectation values of the populations follow a simple rate equation. This means that the qubit population p_q should be governed by a linear differential equation of the form

$$\dot{p}_\mathrm{q}(t) = -\Gamma_1(t) \cdot [p_\mathrm{q}(t) - p_\mathrm{eq}(t)] \tag{3.2}$$

with potentially time-dependent coefficients $\Gamma_1(t)$, the qubit relaxation rate, and $p_\mathrm{eq}(t)$, the equilibrium population, i.e. the qubit population at which $\dot{p}_\mathrm{q}(t)$ would vanish. In contrast to the time-convolutionless master equation [122], the coefficients $\Gamma_1(t)$ and $p_\mathrm{eq}(t)$ are given by the current TLS populations, which in turn might depend on the qubit history. The main task is therefore to validate Eq. 3.2 and to derive exact or approximate solutions for different scenarios, which in retrospect allow to draw conclusions about the TLS environment.

It should be noted that by observing the qubit population dynamics alone, $\Gamma_1(t)$ and $p_\mathrm{eq}(t)$ are a priori unknown functions that can not be disentangled. However, if one has access to both qubit transition rates $\Gamma_{\uparrow,\downarrow}$, for instance by resetting the qubit to its ground or excited state and measuring the subsequent population change, the time-dependent transition rates are given by

$$\Gamma_\uparrow(t) = \dot{p}_\mathrm{q}(t)\big|_{p_\mathrm{q}=0} \quad \text{and} \quad \Gamma_\downarrow(t) = -\dot{p}_\mathrm{q}(t)\big|_{p_\mathrm{q}=1} \tag{3.3}$$

after resetting to the ground or excited state, respectively. This method is applicable to any qubit with active state reset capability. With Eq. 3.2, one then obtains the usual expressions

$$\Gamma_1(t) = \Gamma_\uparrow(t) + \Gamma_\downarrow(t), \tag{3.4}$$

$$p_{\mathrm{eq}}(t) = \Gamma_\uparrow(t)/\Gamma_1(t). \tag{3.5}$$

For a TLS environment it turns out that Γ_1 is constant over time, i.e. it does not depend on the TLS populations, which also implies that Γ_1 is temperature-independent (cf. Eqs. 3.48 and 3.69 and Fig. 2.20b).

3.2 Qubit and a single TLS

In view of the experimental findings of long-lived TLSs and a qubit decoherence that exceeds the qubit relaxation by an order of magnitude, we will first discuss the cross-relaxation of the qubit and a single TLS (Fig. 3.2).

Figure 3.2: Cross-relaxation between the qubit and a single TLS. Adapted from Ref. [21].

3.2.1 Stochastic Liouville equation

The principle of cross-relaxation can be best understood in the semi-classical framework of the stochastic Liouville equation. Time-dependent stochastic Hamiltonians have first been introduced by Anderson and Kubo to explain the motional narrowing of line-shapes in NMR [123, 124]. To allow for a simple and instructive analysis, the qubit-TLS system is assumed to be lossless. In this case, the environmental perturbation of the qubit and the TLS can be thought of as a classical longitudinal noise source. The time-dependent Hamiltonian reads

$$H_{\mathrm{tot}} = \frac{\hbar\omega_{\mathrm{q}}}{2}\,\sigma_{\mathrm{q}}^z + \frac{\hbar\omega_{\mathrm{t}}}{2}\,\sigma_{\mathrm{t}}^z + \hbar g\,\sigma_{\mathrm{q}}^x\sigma_{\mathrm{t}}^x + \hbar\eta_{\mathrm{q}}(t)\sigma_{\mathrm{q}}^z + \hbar\eta_{\mathrm{t}}(t)\sigma_{\mathrm{t}}^z. \tag{3.6}$$

Here, ω_{q} and ω_{t} are the qubit and TLSs frequencies, respectively, g is the coupling strength and $\eta_{\mathrm{q,t}}$ are classical noise sources with a certain power spectral density and zero mean. The time-dependent system wave function

$$\psi(t) = \sum c_k(t)\,|v_k\rangle \tag{3.7}$$

can be expressed in the computational basis $|v_k\rangle$. The wave functions $|v_0\rangle$, $|v_1\rangle$, $|v_2\rangle$, $|v_3\rangle$ correspond to the states $|11\rangle$, $|01\rangle$, $|10\rangle$, $|00\rangle$, respectively, with the first entry for the qubit and the second for the TLS. For more details on the notation see

App. B. With respect to this basis, the matrix elements $\mathbf{H}_{mn} = \langle v_m | H | v_n \rangle$ yield the time-dependent Hamiltonian in matrix form

$$\mathbf{H}_{\text{tot}} = \hbar \begin{pmatrix} \frac{\sigma}{2} & & & g \\ & \frac{\delta}{2} & g & \\ & g & -\frac{\delta}{2} & \\ g & & & -\frac{\sigma}{2} \end{pmatrix} + \hbar \begin{pmatrix} \eta_\sigma(t) & & & \\ & \eta_\delta(t) & & \\ & & -\eta_\delta(t) & \\ & & & -\eta_\sigma(t) \end{pmatrix}. \tag{3.8}$$

Here, the sum frequency $\sigma = \omega_{\text{q}} + \omega_{\text{t}}$, the difference frequency $\delta = \omega_{\text{t}} - \omega_{\text{q}}$, and the mutual dephasings $\eta_\delta(t) = \eta_{\text{t}}(t) - \eta_{\text{q}}(t)$ and $\eta_\sigma(t) = \eta_{\text{q}}(t) + \eta_{\text{t}}(t)$ have been introduced. The Hamiltonian can be decomposed into two manifolds, namely the single- and the two-photon manifold. In the former, the excitation is conserved and undergoes flip-flop processes (zero quantum transitions), while in the latter the excitation undergoes flip-flip processes (double quantum transitions). Since the derivation is analogous in both manifolds, we continue in the single-photon manifold, which gives the main contribution to the population transfer. The Hamiltonian of the single-photon manifold can be written as

$$H = H_0 + H_\eta(t) = \frac{\hbar\delta}{2}\sigma_z + \hbar g \sigma_x + \hbar\eta(t)\sigma_z \tag{3.9}$$

using the shorter notation $\eta := \eta_\delta$. The evolution of the density matrix is given by the Liouville-von Neumann equation

$$\dot{\rho} = -\frac{i}{\hbar}[H_0 + H_\eta(t), \rho]. \tag{3.10}$$

In superoperator notation the equation is of the form

$$\dot{\rho} = (\mathcal{L}_0 + \mathcal{L}_\eta(t))\rho = \mathcal{L}(t)\rho, \tag{3.11}$$

where $\mathcal{L}_0$ and $\mathcal{L}_\eta$ are denoted as Liouvillian superoperators. These superoperators may as well be represented in matrix form acting on the density matrix that must be cast in form of a vector, $\rho = (|c_1|^2, c_2c_1^*, c_1c_2^*, |c_2|^2)^T$. For more details see App. B. To give an example, the Liouvillian for the free evolution is of the form

$$\mathcal{L}_0 = i \begin{pmatrix} 0 & -g & g & 0 \\ -g & \delta & 0 & g \\ g & 0 & -\delta & -g \\ 0 & g & -g & 0 \end{pmatrix}. \tag{3.12}$$

Since $\mathcal{L}_0$ and $\mathcal{L}_\eta(t)$ do not commute, the Liouville-von Neumann equation can not be solved by a simply matrix exponential. Nevertheless, it has the formal solution $\rho(t) = \Phi(t - t_0)\rho(t_0)$ with $\Phi(t)$ being the propagator.

A first insight into the stochastic dynamics can be obtained when the noise $\eta(t)$ is considered as a Poisson process, meaning that $\eta(t)$ is non-zero only for very short random moments in time during which it reaches large amplitudes. In the limit of vanishing interaction time the noise can be seen as a trace of δ-functions appearing on average with the rate λ. The free evolution, encoded in $\mathcal{L}_0$ can be neglected during the interaction, and integration of $\mathcal{L}_\eta(t)$ yields the propagator

$$\Phi_\varphi = \begin{pmatrix} 1 & & & \\ & e^{i\varphi} & & \\ & & e^{-i\varphi} & \\ & & & 1 \end{pmatrix}, \tag{3.13}$$

describing jumps in phase with the random variable φ that belongs to a probability distribution $P(\varphi)$. Without loss of generality, the distribution $P(\varphi)$ can be assumed to be of zero mean.

A single realization of this stochastic process is characterized by the set of jumps $\{t_1, \varphi_1; \ldots; t_n, \varphi_n\}$ within the time t, which yields the density matrix

$$\rho(t) = \Phi_0(t - t_n)\Phi_{\varphi_n} \ldots \Phi_0(t_2 - t_1)\Phi_{\varphi_1}\Phi_0(t_1 - t_0)\rho(t_0). \tag{3.14}$$

Here, $\Phi_0(t) = e^{\mathcal{L}_0 t}$ is the propagator of the unperturbed system, which is of the form

$$\Phi_0(t) = \mathbf{S}\begin{pmatrix} 1 & & & \\ & e^{i\Omega t} & & \\ & & e^{-i\Omega t} & \\ & & & 1 \end{pmatrix}\mathbf{S}^T, \text{ with } \mathbf{S} = \begin{pmatrix} \frac{1}{\sqrt{2}} & \frac{2g}{2\Omega} & \frac{2g}{2\Omega} & \frac{\delta}{\sqrt{2}\Omega} \\ 0 & -\frac{\Omega+\delta}{2\Omega} & \frac{\Omega-\delta}{2\Omega} & \frac{2g}{\sqrt{2}\Omega} \\ 0 & \frac{\Omega-\delta}{2\Omega} & -\frac{\Omega+\delta}{2\Omega} & \frac{2g}{\sqrt{2}\Omega} \\ \frac{1}{\sqrt{2}} & -\frac{2g}{2\Omega} & -\frac{2g}{2\Omega} & \frac{\delta}{\sqrt{2}\Omega} \end{pmatrix}. \tag{3.15}$$

This propagator describes vacuum Rabi oscillations between the qubit and the TLS with the Rabi frequency $\Omega = \sqrt{\delta^2 + 4g^2}$. Fig. 3.3 depicts trajectories for strong and seldom events as well as for weak and frequent events. In both cases a stochastic population transfer between the qubit and the TLS is visible. It can be seen that the state vector of the single photon manifold essentially performs a random walk over the excitation Bloch sphere.

In experiments, one often only has access to the statistical density matrix $\rho(t) = \langle\Phi(t - t_0)\rangle\rho(t_0)$, averaged over many individual trajectories. We therefore wish to find an effective propagator $\Phi_{\mathrm{eff}}(t - t_0) = e^{\mathcal{L}_{\mathrm{eff}}(t-t_0)} \approx \langle\Phi(t - t_0)\rangle$ that is local in time. For random jumps in phase this averaging can be performed exactly (s. App. C) and the dynamics is governed by

$$\dot{\rho} = \mathcal{L}_{\mathrm{eff}}\rho = (\mathcal{L}_0 - \lambda(1 - \langle\Phi_\varphi\rangle))\rho. \tag{3.16}$$

The diagonal form of Φ_φ allows us to introduce the mutual dephasing as $\Gamma_2 = \lambda(1 - \langle e^{i\varphi}\rangle)$. Assuming for simplicity a Gaussian distribution for the phase with variance σ^2, one finds

$$\Gamma_2 = \lambda(1 - e^{-\frac{\sigma^2}{2}}) \tag{3.17}$$

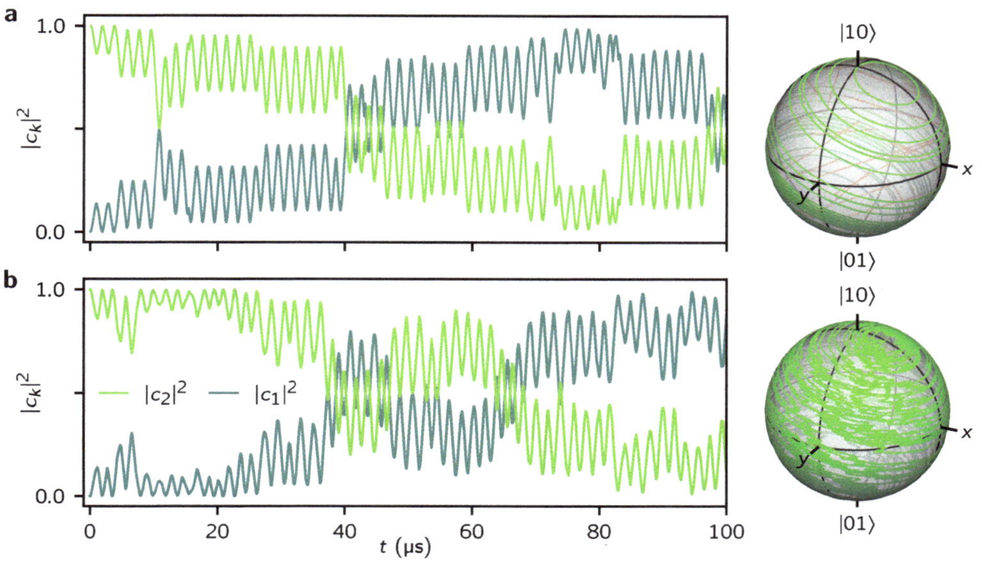

Figure 3.3: Stochastic cross-relaxation. Vacuum Rabi oscillations between the qubit and TLS in the single photon manifold under the presence of random jumps in phase for $\delta = 2\pi \cdot 0.5\,\text{MHz}$ and $g = 2\pi \cdot 0.1\,\text{MHz}$. To the left the qubit (light green) and TLS population (dark green) is shown as a function of time and to the right the stochastic state trajectory is depicted on the Bloch sphere of the single photon manifold. **a** Trajectory for $\lambda = 0.2\,\text{MHz}$ and a Gaussian distribution for $P(\varphi)$ with $\sigma = 2\pi \cdot 0.2$. The red lines in the Bloch sphere indicate the jumps. **b** Trajectory for $\lambda = 2\,\text{GHz}$ and $\sigma = 2\pi \cdot 0.002$. This large number of weak events renders this trajectory a solution of the stochastic Schrödinger equation.

and conseqeuently

$$\sigma = \sqrt{2 \ln \frac{\lambda}{\lambda - \Gamma_2}} \propto \frac{1}{\sqrt{\lambda}} \quad \text{for } \lambda \gg 0 \tag{3.18}$$

with the typical scaling of the Wiener process as λ approaches infinity.

At this point, it should be emphasized that the derivation of jumps in phase are highly idealized and unphysical since infinite energies are involved. A physically motivated modeling of the noise can be achieved with the Kubo-Anderson process [123–126]. This process was first conceived to explain the motional narrowing of line shapes in NMR, which undergo a transition from a Gaussian to a narrowed Lorentzian line shape with increasing temperature and hence motion in the system. However, as long as the noise is finite, the tails of the line shape will fall off much faster than predicted by the Lorentzian line shape [125]. The Lorentzian line shape may therefore be seen as the white noise limit, as will be shown explicitly in Sec. 3.3. The idea of the Kubo-Anderson process is to consider a Poisson process of jumps in frequency according to some frequency distribution. For the simple Kubo oscillator the averaging of the propagator can be obtained analytically [126]. Here, the qubit and TLS averaging is much more complicated due to the off-diagonal coupling term [127]. Alternatively, one may approximate the stochastic propagator $\langle \Phi(t) \rangle$, which is discussed in great detail in Ref. [125]. We present thereof the relevant derivation and conditions that lead to a time-local master equation.

In short summary, the effective Liouvillian $\mathcal{L}_{\text{eff}}$ gives a good approximation when the noise is sufficiently weak and incoherent, which allows us to integrate the noise independently on short time scales. For our system we can already surmise that for the correlation time τ_c of the noise it must hold $\tau_c \ll 2\pi/\Omega$. It follows that the coherence entries of the density matrix accumulate an additional random phase φ within the correlation time. For isotropic noise, one then obtains the decoherence rate $\Gamma_2 = (1 - \langle e^{i\varphi} \rangle)/\tau_c$, in congruence with the previous derivation.

More explicitly, $\mathcal{L}_{\text{eff}}$ can be obtained approximately in the interaction picture using second-order perturbation theory [125]. The density matrix in the interaction picture is defined by $\rho_{\text{I}} = e^{-\mathcal{L}_0 t} \rho$ and governed by the equation of motion

$$\dot{\rho}_{\text{I}} = e^{-\mathcal{L}_0 t} \mathcal{L}_\eta(t) e^{\mathcal{L}_0 t} \rho_{\text{I}} = \mathcal{L}_{\text{I}}(t) \rho_{\text{I}}. \tag{3.19}$$

The formal solution $\rho_{\text{I}}(t) = U(t, t_0) \rho_{\text{I}}(t_0)$ is given by the Dyson series

$$U(t, t_0) = 1 + \int_{t_0}^{t} \mathrm{d}t_1 \mathcal{L}_{\text{I}}(t_1) + \int_{t_0}^{t} \mathrm{d}t_1 \mathcal{L}_{\text{I}}(t_1) \int_{t_0}^{t_1} \mathrm{d}t_2 \mathcal{L}_{\text{I}}(t_2) + \dots. \tag{3.20}$$

It follows that the evolution of the statistical density matrix in the Schrödinger picture is given by

$$\begin{aligned}
\dot{\rho} &= \mathcal{L}_0 \rho + e^{\mathcal{L}_0 t} \langle \dot{\rho}_{\text{I}} \rangle \\
&= \mathcal{L}_0 \rho + e^{\mathcal{L}_0 t} \frac{\mathrm{d}}{\mathrm{d}t} \langle U(t, t_0) \rangle \, \rho_{\text{I}}(t_0) \\
&= \mathcal{L}_0 \rho + e^{\mathcal{L}_0 t} \underbrace{\langle e^{-\mathcal{L}_0 t} \mathcal{L}_\eta(t) e^{\mathcal{L}_0 t} \rangle}_{=0} \rho_{\text{I}}(t_0) \\
&\quad + \int_{t_0}^{t} \mathrm{d}t' \langle \mathcal{L}_\eta(t) \underbrace{e^{-\mathcal{L}_0(t-t')}}_{\approx 1} \mathcal{L}_\eta(t') \underbrace{e^{-\mathcal{L}_0(t-t')}}_{\approx 1} \rangle \cdot e^{\mathcal{L}_0 t} \underbrace{\rho_{\text{I}}(t_0)}_{\approx \rho_{\text{I}}(t)} + \dots \\
&\approx \left(\mathcal{L}_0 + \frac{1}{2} \int_{-\infty}^{\infty} \langle \mathcal{L}_\eta(0) \mathcal{L}_\eta(\tau) \rangle \mathrm{d}\tau \right) \rho = \mathcal{L}_{\text{eff}} \, \rho.
\end{aligned} \tag{3.21}$$

Several assumptions have been made above. First, it is assumed that the noise is isotropic. Next, the noise is assumed to be sufficiently incoherent such that $\tau_c \ll 2\pi/\Omega$ and consequently $t_0 \approx t - \tau_c$. Furthermore, it is assumed that the noise is weak enough so that $\rho_{\text{I}}(t_0)$ can be approximated by its current time $\rho_{\text{I}}(t)$ and the Dyson series can be terminated at the second order. These assumptions may

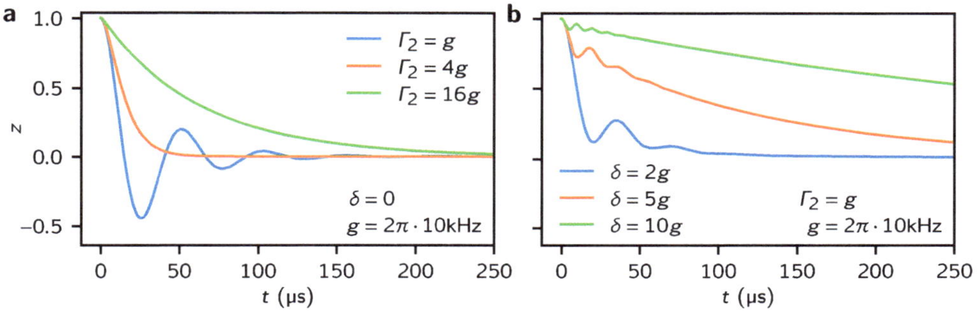

Figure 3.4: Qubit and TLS cross-relaxation. a, b Relaxation of the polarization z in the single-photon manifold of the qubit-TLS system for various decoherence rates Γ_2 (panel a) and frequency detunings δ (panel b). The coupling strength g was chosen similar to the experiment (Sec. 2.8). Reproduced from Ref. [86].

be summarized as the narrowing condition [125]. Further details can be found in Ref. [128]. Finally, the noise is assumed to be stationary, which allows to extend the integral boundaries to infinity and links the decoherence rate $\Gamma_2 = 2S_{\eta\eta}(0)$ with the power spectral density $S_{\eta\eta}(\omega)$ of the noise $\eta(t)$ [129]. The effective Liouvillian derived here coincides with the one in Eq. 3.16.

Since the population $(\rho_{11} + \rho_{22}) \in [0, 1]$ in the single-photon manifold is conserved, Eq. 3.21 is a special case of the Bloch equations [130, 131]. Defining the polarization $z = \rho_{11} - \rho_{22}$, and directions $x = 2 \cdot \mathrm{Re}(\rho_{21})$, and $y = 2 \cdot \mathrm{Im}(\rho_{21})$, the Liouville-von Neumann equation is of the form

$$\frac{\mathrm{d}}{\mathrm{d}t}\begin{pmatrix} x \\ y \\ z \end{pmatrix} = -\begin{pmatrix} \Gamma_2 & \delta & 0 \\ -\delta & \Gamma_2 & 2g \\ 0 & -2g & 0 \end{pmatrix} \cdot \begin{pmatrix} x \\ y \\ z \end{pmatrix}. \tag{3.22}$$

It immediately follows from the steady-state solution that, on average, both states of the single-photon manifold are equally populated. For a finite temperature, this steady-state solution contradicts the Boltzmann distribution when the qubit and the TLS have different energies. This result is typical for stochastic Liouville equations and a consequence of the real valued noise acting as an infinite temperature environment [127]. The steady state solution is a good approximation as long as the energy difference is small in comparison to the temperature. In Fig. 3.4 various solutions of Eq. 3.22 are depicted. The computation is discussed in the next section.

3.2.2 Coherent and incoherent cross-relaxation

The cross-relaxation between the qubit and the TLS is composed of coherent relaxation in form of damped vacuum Rabi oscillations and incoherent relaxation, see

Fig. 3.4. This section is devoted to quantifying their contributions with a focus on the condition that allows to neglect the coherent population transfer.

Solving Eq. 3.22 requires finding the roots of the characteristic polynomial, which is cubic and can not easily be factored. Fortunately, the structure of the characteristic polynomial of $\mathcal{L}_{\text{eff}}$ in Eq. 3.22 allows for a good approximation in a wide parameter range. It holds

$$\lambda^3 - 2\Gamma_2\lambda^2 + (\Gamma_2^2 + \delta^2 + 4g^2)\lambda - 4g^2\Gamma_2 = 0, \tag{3.23}$$

and by defining $\mu = \lambda/\Gamma_2$ we obtain

$$\mu^3 - 2\mu^2 + \frac{\Gamma_2^2 + \delta^2 + 4g^2}{\Gamma_2^2}\mu - 4\frac{g^2}{\Gamma_2^2} = 0. \tag{3.24}$$

Now, let μ_i denote the roots of the above polynomial and let μ_0 denote the real root with lowest value in case that all roots are real. For the corresponding eigenvectors one finds $v_i = \left(\delta\mu_i/(1-\mu_i), \quad -\Gamma_2\mu_i, \quad 2g\right)^T$, which do not form an orthogonal basis since $\mathcal{L}_{\text{eff}}$ is not a normal operator. Nevertheless, when $\mu_0 \to 0$ one can already surmise from the form of the eigenvectors that this eigenvalue contributes dominantly to the relaxation. Unfortunately, it is not straightforward to give a quantitative answer, particularly because the eigenvalues can become degenerate. It is therefore very instructive to derive a few properties of the roots that allow for a better understanding of the resulting relaxation dynamics.

When $\Gamma_2^2 + \delta^2 \gg 4g^2$ or, equivalently, when $\mu_0 \to 0$, the root μ_0 can be approximated to first order,

$$\mu_0 \approx \frac{4g^2}{\Gamma_2^2 + \delta^2 + 4g^2}. \tag{3.25}$$

Besides the negligible term $4g^2$ in the denominator and an overall factor of 2, we obtain once again the cross-relaxation rate (cf. Eq. 3.61). The factor of 2 in the numerator simply results from the fact that λ_0 describes the qubit and TLS relaxation, whereas the cross-relaxation rate Γ_{qt} is a transition rate between the qubit and TLS.

The next two roots $\mu_{1,2}$ can in principle be expressed with the help of μ_0. However, this requires a case analysis on the sign of the discriminant D of the third-order polynomial and furthermore no intuition for the behavior of the roots is gained. Instead, a different approach is presented. First, using the Routh-Hurwitz stability criterion, one can easily show that $\text{Re}(\mu_i) \in (0, 1]$. Second, we have $\sum_i \mu_i = 2$. Thus, whenever $\mu_0 \to 0$ we have that $\text{Re}(\mu_{1,2}) \to 1$, and if the discriminant[1] $D > 0$, the imaginary part is given by $\text{Im}(\mu_{1,2}) = \pm\sqrt{(\delta^2 + 4g^2)/\Gamma_2^2 - \mu_0 + 3/4\mu_0^2} \to \pm\sqrt{\delta^2 + 4g^2}/\Gamma_2$. As a

[1]Note that there are different sign conventions in the definition of the discriminant.

side note, the value $\mu_i = 1$ is realized for one root if and only if $\delta = 0$. In this special case, the polynomial decomposes into $(\mu - 1)(\mu^2 - \mu + 4g^2/\Gamma_2^2) = 0$ with solutions

$$\mu_i = 1 \quad \vee \quad \mu_{j,k} = \frac{1}{2} \pm \sqrt{\frac{1}{4} - \frac{4g^2}{\Gamma_2^2}}. \tag{3.26}$$

With this understanding of the roots, it becomes advantageous to express the relaxation dynamics with the help of the roots. The relaxation dynamics, following an initial condition x_0, y_0, and z_0, can be expressed as

$$z(t) = \sum_{i=0}^{2} (1 - \mu_i) \frac{(1 - \mu_{i+1})(1 - \mu_{i+2}) 2g\Gamma_2\, x_0 + 2g\delta\, y_0 + \mu_{i+1}\mu_{i+2}\, \delta\Gamma_2\, z_0}{(\mu_{i+1} - \mu_i)(\mu_{i+1} - \mu_i)\delta\Gamma_2} \cdot e^{-\mu_i \Gamma_2 t}, \tag{3.27}$$

where we defined $\mu_{i+3} = \mu_i$ cyclically. More compactly one may write

$$z(t) = \sum_{i=0}^{2} (u_i x_0 + v_i y_0 + w_i z_0) e^{-\mu_i \Gamma_2 t} = \sum_{i=0}^{2} c_i e^{-\mu_i \Gamma_2 t}. \tag{3.28}$$

Next, it must be shown that $c_0 \to z_0$ for $\mu_0 \to 0$. In this case, after approximately the decoherence time $1/\Gamma_2$, when the initial coherent onset has vanished, the remaining term is the slow decay of the polarization from z_0 at the rate $\mu_0 \Gamma_2$. Using Eq. 3.24 and our knowledge of the roots, one finds

$$u_0 = \frac{\sqrt{\mu_0 \mu_1 \mu_2 (1 - \mu_0)(1 - \mu_1)(1 - \mu_2)}}{(\mu_1 - \mu_0)(\mu_2 - \mu_0)} = O(\sqrt{\mu_0}), \tag{3.29}$$

$$v_0 = \frac{\sqrt{\mu_0 \mu_1 \mu_2}}{(\mu_1 - \mu_0)(\mu_2 - \mu_0)} = O(\sqrt{\mu_0}), \tag{3.30}$$

$$w_0 = \frac{(1 - \mu_0)\mu_1 \mu_2}{(\mu_1 - \mu_0)(\mu_2 - \mu_0)} = 1 - O(\mu_0). \tag{3.31}$$

Here, we see here that the incoherent relaxation with $e^{-\mu_0 \Gamma_2 t}$ with the prefactor c_0 is well behaved when $\mu_0 \to 0$.

In order to also be concrete for the coherent population transfer with $e^{-\mu_{1,2}\Gamma_2 t}$, the prefactors $c_{1,2}$ must be handled, which may diverge when the eigenvalues become degenerate. One way to circumvent the divergence is to calculate the area A under the relaxation curve by integrating Eq. 3.28, which can then be compared to the corresponding exponential decay with the rate $\mu_0 \Gamma_2$. The area simplifies to

$$\begin{aligned} A(x_0, y_0, z_0) &= \int_0^\infty z(t)\mathrm{d}t = \int_0^\infty \sum_{i=0}^{2} c_i e^{-\mu_i \Gamma_2 t} \mathrm{d}t \\ &= \frac{\delta x_0 + \Gamma_2 y_0}{2g\Gamma_2} + \frac{\Gamma_2^2 + \delta^2}{4g^2 \Gamma_2} z_0. \end{aligned} \tag{3.32}$$

Interestingly, the inverse of the prefactor of z_0 is exactly the two-fold cross-relaxation rate Γ_{qt} (Eq. 3.61). It can therefore be concluded that Γ_{qt} is the average rate when the system starts in an incoherent state, meaning that $x_0 = y_0 = 0$. There is still a fraction of z_0 that decays coherently, which can be quantified as

$$\begin{aligned} \epsilon &= \left(A(0,0,z_0) - \int_0^\infty z_0 e^{-\mu_0 \Gamma_2 t}\, \mathrm{d}t \right) \Big/ \int_0^\infty z_0 e^{-\mu_0 \Gamma_2 t}\, \mathrm{d}t \\ &= \mu_0 \left(\frac{\mu_1 + \mu_2}{\mu_1 \mu_2} - 1 \right), \end{aligned} \tag{3.33}$$

approaching zero for $\mu_0 \longrightarrow 0$. Similarly, one might wonder to what extend an initial coherence is converted into polarization. With $z_{\mathrm{max}} = \rho_{11} + \rho_{22}$ the maximal area is found to be

$$\max_{x_0, y_0} A(x_0, y_0, z_0) = \frac{\sqrt{z_{\mathrm{max}}^2 - z_0^2}\sqrt{\Gamma_2^2 + \delta^2}}{2g\Gamma_2}. \tag{3.34}$$

This area should be compared to $\mu_0 \Gamma_2$ to give the effectively reached polarization

$$z_{\mathrm{eff}} = \mu_0 \frac{\sqrt{z_{\mathrm{max}}^2 - z_0^2}\sqrt{\Gamma_2^2 + \delta^2}}{2g}, \tag{3.35}$$

which approaches zero for $\mu_0 \longrightarrow 0$.

In Fig. 3.5 the relative error ϵ of neglecting the coherent cross-relaxation is shown as a function of the mutual decoherence and detuning. In summary, we conclude from Eq. 3.33 that if $\sqrt{\Gamma_2^2 + \delta^2} > 4g$ and equivalently when $\mu_0 \longrightarrow 0$, the relaxation of the population is governed by

$$\dot{z} = -\mu_0 \Gamma_2 z. \tag{3.36}$$

Solomon equations

For sake of completeness, the current scenario of longitudinal noise allows for a convenient derivation of the Solomon equation. From Eq. 3.36, it follows for the populations in the single photon manifold

$$\dot{\rho}_{22} = -\dot{\rho}_{11} = -\mu_0 \Gamma_2 \frac{\rho_{22} - \rho_{11}}{2}. \tag{3.37}$$

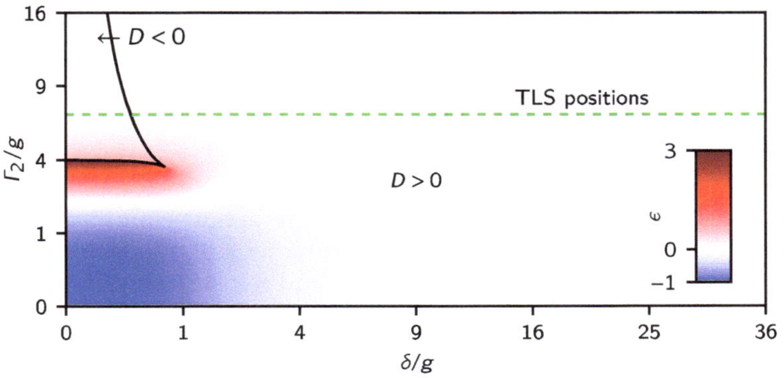

Figure 3.5: Fraction of coherent cross-relaxation. Relative proportion ϵ of the initial polarization that does not relax via the incoherent cross-relaxation rate, as a function of the detuning and decoherence. Note the quadratic axes. The black curve shows where the discriminant D is zero. For $D > 0$ the polarization can show oscillatory behavior. In the region where $\sqrt{\Gamma_2^2 + \delta^2} > 4g$, the incoherent cross-relaxation becomes the dominant relaxation mechanism. For reference, the green dashed line shows where the TLSs are located in the experiment assuming a mutual decoherence of $\Gamma_2 = 0.5\,\text{MHz}$ (Sec. 2.8). Reproduced from Ref. [86].

Insertion of the population expectation values $p_\text{q} = \rho_{00} + \rho_{22}$ and $p_\text{t} = \rho_{00} + \rho_{11}$ and for simplicity neglecting the relaxation in the two photon manifold yields the Solomon equations [83]

$$\dot{p}_\text{q} = -\frac{\mu_0\Gamma_2}{2}p_\text{q} + \frac{\mu_0\Gamma_2}{2}p_\text{t}, \tag{3.38}$$

$$\dot{p}_\text{t} = -\frac{\mu_0\Gamma_2}{2}p_\text{t} + \frac{\mu_0\Gamma_2}{2}p_\text{q}, \tag{3.39}$$

with the incoherent cross-relaxation rate [83, 84, 127]

$$\Gamma_\text{qt} = \frac{\mu_0\Gamma_2}{2} \approx \frac{2g^2\Gamma_2}{\Gamma_2^2 + \delta^2}. \tag{3.40}$$

3.3 Bloch-Redfield equation

For a later comparison with the Solomon equations, we briefly review the opposite limit, where the qubit is weakly coupled to a large number of TLSs such that they form a Born-Markovian environment for the qubit. As a result, the qubit will experience a quantum mechanical noise source. The reduced density matrix of the qubit is governed by the Bloch-Redfield master equation, which was first put forward by Wangsness and Bloch [132, 133] and later refined by Redfield for the off-diagonal entries of the density matrix [134]. The Bloch-Redfield master equation rests on the

Born approximation and the first and second Markov approximations. Typically, the secular approximation is used in addition to yield a master equation of Lindblad type. With the secular approximation the derivation is analogous to the quantum optical master equation that describes the interaction of a system with a bosonic environment [122].

In the interaction picture, the density matrix of the total system is governed by $\dot{\rho}_{\mathrm{tot}} = -\frac{i}{\hbar}\left[H_{\mathrm{int}}(t), \rho_{\mathrm{tot}}\right]$, with H_{int} denoting the interaction Hamiltonian of Eq. 3.1. Unlike in Sec. 3.2, the superscript I denoting the interaction picture is omitted here for clarity. Formal integration of this equation and reinsertion yields

$$\dot{\rho}_{\mathrm{tot}}(t) = -\frac{i}{\hbar}[H_{\mathrm{int}}(t), \rho_{\mathrm{tot}}(t_0)] - \frac{1}{\hbar^2}\int\limits_{t_0}^{t}[H_{\mathrm{int}}(t), [H_{\mathrm{int}}(t'), \rho_{\mathrm{tot}}(t')]]\mathrm{d}t'. \tag{3.41}$$

At this point, the equation is still exact. However, when the Markov approximations are applied, it turns out that a more accurate strategy would have been to use the correct second order approximation of the Dyson series, essentially as used in Eq. 3.21. This less stringent derivation has already been emphasized by Redfield [134] and is known as coarse graining. It does not require the two Markov approximations and the secular approximation. Its full potential has been recognized later by Schaller [135].

In proceeding forward, the Born approximation has to be applied. It is a rather obscure and crude approximation, which entails several assumptions that are shortly discussed in the following. As we are interested in the dynamics of the qubit's reduced density matrix ρ_{q}, we can trace out the TLS bath in Eq. 3.41. However, a dynamical equation for ρ_{q} is only obtained if the qubit and TLS bath are uncorrelated, i.e. $\rho_{\mathrm{tot}} = \rho_{\mathrm{q}} \otimes \rho_{\mathrm{TLSs}}$. It is clear that the system and the bath will become entangled after a short time, which renders this ansatz invalid on the long run. We may however imagine that the system is constantly reinitialized in a product state. A meaningful solution can only be expected when the bath state is chosen at random for the reinitialization. The density matrices of the qubit and TLS bath should therefore be thought of as a statistical mixture. The density matrix of the TLS bath is typically assumed to be diagonal, which is also known as the random phase assumption [134]. In order to obtain a solution that is valid on long time scales, it is further assumed that the bath is in an equilibrium state $\bar{\rho}_{\mathrm{TLSs}}$ that does not evolve in time. This also means that the TLS populations should not be altered significantly by the qubit, which is justified when the qubit interacts weakly with a large number of TLSs. The Born approximation is summarized as follows:

$$\dot{\rho}_{\mathrm{q}}(t) = -\frac{1}{\hbar^2}\int\limits_{t_0}^{t}\mathrm{Tr}\{[H_{\mathrm{int}}(t), [H_{\mathrm{int}}(t'), \rho_{\mathrm{q}}(t') \otimes \bar{\rho}_{\mathrm{TLSs}}]]\}\mathrm{d}t'. \tag{3.42}$$

The first term on the right hand side of Eq. 3.41 vanishes, since $\bar{\rho}_{\mathrm{TLSs}}$ is assumed to be diagonal. For a small number of TLSs with a long intrinsic lifetime, the Born approximation is certainly not fulfilled, requiring a description with the Solomon equations (s. Sec. 3.4). In addition, the Solomon equations remain valid when the qubit and the TLSs are entangled.

Next, $\tau = t - t'$ is introduced and the first and second Markov approximations are applied, i.e. evaluating the qubit's density matrix at the current time t and extending the integration over the bath correlation function $C(\tau)$ to infinity. We then obtain

$$\dot{\rho}_{\mathrm{q}}(t) = -\frac{1}{\hbar^2} \int_0^\infty C(\tau)[\sigma_{\mathrm{q}}^x(t), \sigma_{\mathrm{q}}^x(t-\tau)\rho_{\mathrm{q}}(t)] + C^*(\tau)[\rho_{\mathrm{q}}(t)\sigma_{\mathrm{q}}^x(t-\tau), \sigma_{\mathrm{q}}^x(t)]\,\mathrm{d}\tau. \quad (3.43)$$

With the secular approximation, which means removing the explicit time-dependence, the above equation becomes the Bloch-Redfield master equation, which is of Lindblad type. The bath correlation function is given by

$$C(\tau) = \hbar^2 \sum_k g_k^2 \,\mathrm{Tr}\{\sigma_k^x(t)\sigma_k^x(t-\tau)\bar{\rho}_{\mathrm{TLSs}}\} \quad (3.44)$$

$$= \hbar^2 \sum_k g_k^2 \left[p_{\mathrm{t}}^k e^{i\omega_k\tau} + (1-p_{\mathrm{t}}^k)e^{-i\omega_k\tau}\right]. \quad (3.45)$$

The density matrix of the TLS bath contains the populations p_{t}^k of the TLSs. Typically, the TLSs are assumed to be in thermal equilibrium and given by the Fermi-Dirac distribution $p_{\mathrm{t}}^k = n(\omega_k)$. As a consequence, the qubit will relax to the same thermal equilibrium with $p_{\mathrm{q}} = n(\omega_q)$.

The usual way to continue the derivation from Eq. 3.45 is to consider the limit where the TLSs become dense in frequency with a vanishing coupling strength such that they form a spectral coupling density $\gamma(\omega) = \langle g^2\nu_{\mathrm{t}}\rangle(\omega)$, with ν_{t} being the TLS density [136, 137]. However, we will see in Sec. 3.4 and Sec. 3.6 that the sheer amount of TLSs does not necessarily lead to a Born-Markovian environment. The transition to a density should rather be viewed as an ensemble average over qubit and TLS frequencies for many instances of the experiment. In Fig. 3.6a we illustrate $\gamma(\omega)$ and the qubit's density of states ν_{q}. The correlation function can now be expressed via the integral

$$C(\tau) = \hbar^2 \int_0^\infty \gamma(\omega)\left[p_{\mathrm{t}}(\omega)e^{i\omega\tau} + (1-p_{\mathrm{t}}(\omega))e^{-i\omega\tau}\right]\mathrm{d}\omega \quad (3.46)$$

$$= \hbar^2 \int_{-\infty}^\infty [s(\omega) + a(\omega)]e^{i\omega\tau}\mathrm{d}\omega, \quad (3.47)$$

where we introduced the symmetric $s(\omega) = s(-\omega)$ and antisymmetric $a(\omega) = -a(-\omega)$ frequency components of the bath correlation function with $s(\omega \geq 0) = \gamma(\omega)/2$ and $a(\omega > 0) = \gamma(\omega)\cdot[p_{\mathrm{t}}(\omega) - 1/2]$, which help to compute the half-sided inverse Fourier transform in Eq. 3.43.

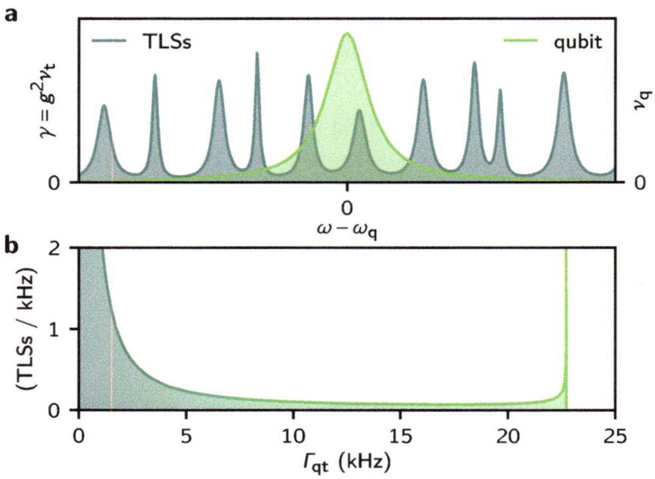

Figure 3.6: Illustrative model for a cross-relaxation distribution. a Spectral coupling density γ of the TLSs and density of states ν_q of the qubit. The TLSs are randomly distributed in frequency with individual coupling strength to the qubit and intrinsic decoherence. **b** Density of cross-relaxation rates following from Eq. 3.50 for a random TLS frequency distribution, assuming the same coupling strength g and mutual decoherence Γ_2 for all TLSs. The parameters are taken from Ref. [21]. If the TLSs have some distribution in g and Γ_2, the divergence at the maximal cross-relaxation rate washes out and disappears, whereas the divergence for low cross-relaxation rates remains. The color gradient from blue to gray illustrates the increasing frequency detuning between the TLSs and the qubit of the corresponding cross-relaxation rates. Reproduced from Ref. [86].

Next, we can rewrite Eq. 3.43 as $\dot{\rho}_q = \mathcal{L}\rho_q$ in superoperator notation. Separating $\mathcal{L}$ in its real and imaginary parts, we define $\dot{\rho}_q = \dot{\rho}_q^{\text{relax}} + \dot{\rho}_q^{\text{Lamb}} = \text{Re}(\mathcal{L}) \cdot \rho_q + i\text{Im}(\mathcal{L}) \cdot \rho_q$. The real-part describes the qubit relaxation and decoherence

$$\frac{\dot{\rho}_q^{\text{relax}}}{2\pi\gamma(\omega_q)} = -\frac{1}{2}\begin{pmatrix} 2\rho_{00} & \rho_{01} \\ \rho_{10} & -2\rho_{00} \end{pmatrix} + p_t(\omega_q)\sigma_z. \tag{3.48}$$

Here, we obtain the important result that the qubit relaxation $\Gamma_1 = 2\pi\gamma(\omega_q)$ and decoherence $\Gamma_2 = \Gamma_1/2$ are independent of the TLS populations [138]. For completeness, the imaginary part incorporates the Lamb shift

$$\dot{\rho}_q^{\text{Lamb}} = i \ \text{P.V.}\int_{-\infty}^{\infty} \frac{2\omega_q s(\omega)}{\omega_q^2 - \omega^2} d\omega \cdot \begin{pmatrix} 0 & -\rho_{01} \\ \rho_{10} & 0 \end{pmatrix}, \tag{3.49}$$

where the integral has to be evaluated by means of the Cauchy principal value. Again, there is no dependence on the TLS populations. Note that by also integrating over negative frequencies, contributions from counter-rotating waves are included in the Lamb shift, and thus, the Lamb shift vanishes for a constant spectral coupling density.

If we had applied the rotating-wave approximation already in the Hamiltonian in Eq. 3.1, the Lamb shift would not contain contributions from counter-rotating waves [136, 139] and for a constant spectral coupling density the Lamb shift would diverge.

In view of the following section, the Bloch-Redfield master equation can also be used to estimate the cross-relaxation rate between the qubit and a single TLS. Assuming the qubit and TLS spectral densities are Lorentz distributions centered at ω_q and $\omega_q + \delta$ with linewidths Γ_2^q and Γ_2^t, respectively (s. Fig. 3.6a), then, for a short time such that the Born approximation remains valid, we can calculate the average qubit relaxation rate using partial fractions

$$\begin{aligned}\Gamma_{qt} &= 2\pi g^2 \int_{\infty}^{\infty} \frac{1}{\pi} \frac{\Gamma_2^t}{(\omega - (\omega_q + \delta))^2 + (\Gamma_2^t)^2} \cdot \frac{1}{\pi} \frac{\Gamma_2^q}{(\omega - \omega_q)^2 + (\Gamma_2^q)^2} \mathrm{d}\omega \\ &= \frac{2g^2 \Gamma_2}{\Gamma_2^2 + \delta^2} \end{aligned} \tag{3.50}$$

with the mutual decoherence $\Gamma_2 = \Gamma_2^q + \Gamma_2^t$. This is exactly the result from Eq. 3.40 that was obtained in Sec. 3.2 and shows that the Lorentzian line shape corresponds to the white noise limit. Expanding the scope to include multiple TLSs, the components of the total loss rate of the qubit $\Gamma_1 = \Gamma_q + \sum_k \Gamma_{qt}^k$ can now be defined. The rate Γ_q accounts for additional losses from other environments. If the TLS detunings δ_k are randomly distributed in frequency, Eq. 3.50 defines the density of the cross-relaxation rates $\nu(\Gamma_{qt})$ depicted in Fig. 3.6b. Note that adding the rates implicitly assumes that higher order coherent effects in the system can be ignored, as discussed in detail in the next section.

3.4 Derivation of the Solomon equations for an arbitrary number of TLSs

For a small number of TLSs, the Born approximation is no longer valid since their populations will be altered by the qubit excitation [140, 141]. Ideally, we want to treat the qubit and the TLSs on equal footing. In experiments, the dissipation required to yield a master equation is naturally provided by the surrounding environment of the qubit and the TLSs. In NMR, this environment is known as the spin-lattice, for which a detailed derivation can be found in Ref. [128]. The spin-lattice can either be modeled as a classical noise source in the framework of the stochastic Liouville equation or treated quantum mechanically, essentially following the Bloch-Redfield formalism. Based on these results, we will construct the Liouvillian from a general Lindblad equation. The Lindblad equation takes over the role of randomization, which was implicitly assumed in the derivation of the Bloch-Redfield master equation within the Born approximation.

The general idea behind the Solomon equations is to reduce the Liouvillian to an equation of motion for the expectation values of the qubit and TLS populations [83, 106]. As will be discussed in more detail in the following, in case the decoherence and frequency spread in the system are large in comparison to the couplings, coherent effects can be neglected. An intermediate step is therefore the derivation of Pauli's master equation for the diagonal components of the density matrix. Here, we use the adiabatic elimination of the coherences. Alternatively, one may derive the Pauli master equation from a generalized Schrieffer-Wolff transformation [142], yet others refer to it as the Nakajima-Zwanzig-Markov-Pauli master equation [143]. However, as already mentioned above, beyond the qubit and a single TLS, deriving closed-form solutions is practically in-feasible due to the emergence of multispin phenomena (cross-correlations), which are discussed in NMR within the framework of magnetization modes [106].

In this context, the main focus is to present a compact derivation of the Solomon equations, which we show to follow from the Markov approximations for the coherences and from the neglect of higher order coherent processes. The derivation is accompanied by showing explicit calculations for the case of a qubit coupled to a single TLS. The Lindblad equation for the Hamiltonian in Eq. 3.1 with individual dissipators for the qubit and each of the TLSs can be expressed as

$$\dot{\rho} = -\frac{i}{\hbar}[H, \rho] + \sum_{\alpha} L_{\alpha} \rho L_{\alpha}^{\dagger} - \frac{1}{2}\{L_{\alpha}^{\dagger} L_{\alpha}, \rho\}. \tag{3.51}$$

Here, L_{α} are the corresponding Lindblad operators. These dissipators are given by $\sqrt{\Gamma_{\downarrow}^{j}} \cdot \sigma_{-}^{j}$, $\sqrt{\Gamma_{\uparrow}^{j}} \cdot \sigma_{+}^{j}$, and $\sqrt{\Gamma_{\varphi}^{j}/2} \cdot \sigma_{z}^{j}$ for relaxation, excitation and dephasing, respectively. The index j is used to denote both the qubit and the TLS variables. Casting the Liouville-von Neumann equation (Eq. 3.51) in superoperator notation $\dot{\rho} = \mathcal{L}\rho$, the Liouvillian $\mathcal{L}$ can be represented by a matrix and ρ by a vector (App. B). In view of Nakajima-Zwanzig's projection operator technique [144], the density matrix $\rho = \left(\rho_{\mathrm{D}}, \rho_{\mathrm{C}}\right)^{T}$ may be sorted for diagonal entries ρ_{D} (populations) and non-diagonal entries ρ_{C} (coherences). Then, Eq. 3.51 reads

$$\dot{\rho} = \begin{pmatrix} \mathbf{\Gamma} & \mathbf{R}^{T} \\ \mathbf{R} & \mathbf{C} \end{pmatrix} \rho. \tag{3.52}$$

The matrix $\mathbf{\Gamma}$ depends only on the decay rates $\Gamma_{\uparrow,\downarrow}^{j}$ and describes the relaxation of the system to thermal equilibrium. The matrix $\mathbf{R}$ couples populations and coherences, giving rise to vacuum Rabi oscillations and energy exchange between the qubit and the TLSs. Therefore, its matrix elements $\mathbf{R}_{mn}$ are either zero or amount to one of the coupling terms, in short $\mathbf{R}_{mn} \in \{0, \pm i g_{k}\}$. The matrix $\mathbf{C}$ describes the oscillations, decoherence and interference of the coherences between the qubit and TLSs as well as between the TLSs. Since the σ^{x}-coupling in Eq. 3.1 induces only one-photon (flip-flop) and two-photon (flip-flip) transitions, we can order the coherences $\rho_{\mathrm{C}} = \left(\rho_{\mathrm{C_Z}}, \quad \rho_{\mathrm{C_D}}, \quad \rho_{\mathrm{C_R}}, \quad \rho_{\mathrm{Q}}\right)^T$ and bring Eq. 3.52 into the form

$$\dot{\rho} = \begin{pmatrix} \boldsymbol{\Gamma} & \mathbf{R}_{\mathrm{Z}}^T & \mathbf{R}_{\mathrm{D}}^T & 0 & 0 \\ \mathbf{R}_{\mathrm{Z}} & \mathbf{C}_{\mathrm{Z}} & 0 & \mathbf{S}_{\mathrm{Z}}^T & 0 \\ \mathbf{R}_{\mathrm{D}} & 0 & \mathbf{C}_{\mathrm{D}} & \mathbf{S}_{\mathrm{D}}^T & 0 \\ 0 & \mathbf{S}_{\mathrm{Z}} & \mathbf{S}_{\mathrm{D}} & \mathbf{C}_{\mathrm{R}} & 0 \\ 0 & 0 & 0 & 0 & \mathbf{Q} \end{pmatrix} \rho \tag{3.53}$$

with the indices Z and D denoting those entries that give rise to zero and double quantum transitions, respectively, i.e. excitation differences. The dynamics of the remaining coherences that do not directly couple to populations are described by $\mathbf{C}_{\mathrm{R}}$ and couple via $\mathbf{S}_{\mathrm{Z,D}}$ to the relevant coherences. Similar to $\mathbf{R}$, one finds $\mathbf{S}_{\mathrm{Z,D};mn} \in \{0, \pm i g_k\}$. All irrelevant coherences ρ_{Q} between even and odd photon states undergo an independent evolution described by $\mathbf{Q}$. The matrix $\mathbf{Q}$ makes up half the size of the Liouvillian. For the qubit and a single TLS, $\mathbf{C}_{\mathrm{R}}$ and $\mathbf{S}_{\mathrm{Z,D}}$ do not exist, and a similar transformation leads to the so-called Redfield kite [106, 145]. The matrix structure of the Liouvillian is depicted in Fig. 3.7 as an illustration for the qubit coupled to one, two, and three TLSs.

We may further decompose $\mathbf{C}_{\mathrm{Z,D}} = \mathbf{D}_{\mathrm{Z,D}} - \mathbf{O}_{\mathrm{Z,D}}$ with $\mathbf{D}$ being diagonal and in charge of the oscillation and decoherence of the relevant coherences. The entries $\mathbf{D}_{\mathrm{Z,D};mm}$, e.g. describing the coherence of the qubit with the kth TLS, are of the form $\mathbf{D}_{\mathrm{Z};mm} = -\Gamma_2^k \pm i\delta_k$ and $\mathbf{D}_{\mathrm{D};mm} = -\Gamma_2^k \pm i\sigma_k$, with the mutual decoherence $\Gamma_2^k = \Gamma_2^{\mathrm{q}} + \Gamma_2^{\mathrm{t}_k}$ and the general detunings $\delta_k = \omega_k - \omega_{\mathrm{q}}$ and $\sigma_k = \omega_{\mathrm{q}} + \omega_k$. The decoherence of the qubit Γ_2^{q} and of the TLSs $\Gamma_2^{\mathrm{t}_k}$ are as usual of the form $\Gamma_2 = \Gamma_\varphi + \Gamma_1/2$ with $\Gamma_1 = \Gamma_\uparrow + \Gamma_\downarrow$. In case the qubit is coupled to only one TLS, the matrix $\mathbf{O}_{\mathrm{Z,D}}$ is zero. For more TLSs, the coherences of different photon manifolds can interfere when the TLSs undergo relaxation processes [146, 147]. Thus, $\mathbf{O}_{\mathrm{Z,D};mn} \in \{0, \pm\Gamma_\uparrow^{\mathrm{t}_k}, \pm\Gamma_\downarrow^{\mathrm{t}_k}\}$ (App. B), and in case the TLSs are lossless, the matrices $\mathbf{C}_{\mathrm{Z,D}}$ are diagonal. For completeness, even though not applicable in the experiment, there are additional off-diagonal matrix elements that arise from vacuum Rabi oscillations in the scenario where the TLSs are coupled with each other. In this case, one has $\mathbf{O}_{\mathrm{Z,D};mn} \in \{0, \pm\Gamma_\uparrow^{\mathrm{t}_k}, \pm\Gamma_\downarrow^{\mathrm{t}_k}, \pm i g_k\}$. Lastly, it is important to mention that the diagonal entries of $\mathbf{C}_{\mathrm{R}}$ always contain various combinations of decoherences and detunings.

In the case of a qubit and a single TLS, one finds for the density matrix entries $\rho_{mn} = \langle v_m|\rho|v_n\rangle$ of the single-photon manifold [84]:

$$\begin{aligned} \dot{\rho}_{11} &= ig(\rho_{12} - \rho_{21}) - (\Gamma_\downarrow^{\mathrm{t}} + \Gamma_\uparrow^{\mathrm{q}})\rho_{11} + \Gamma_\downarrow^{\mathrm{q}}\rho_{00} + \Gamma_\uparrow^{\mathrm{t}}\rho_{33}, \\ \dot{\rho}_{21} &= ig(\rho_{22} - \rho_{11}) - \left(\Gamma_2^{\mathrm{q}} + \Gamma_2^{\mathrm{t}} - i\delta\right)\rho_{21}, \\ \dot{\rho}_{22} &= ig(\rho_{21} - \rho_{12}) - (\Gamma_\downarrow^{\mathrm{q}} + \Gamma_\uparrow^{\mathrm{t}})\rho_{22} + \Gamma_\downarrow^{\mathrm{t}}\rho_{00} + \Gamma_\uparrow^{\mathrm{q}}\rho_{33}, \end{aligned} \tag{3.54}$$

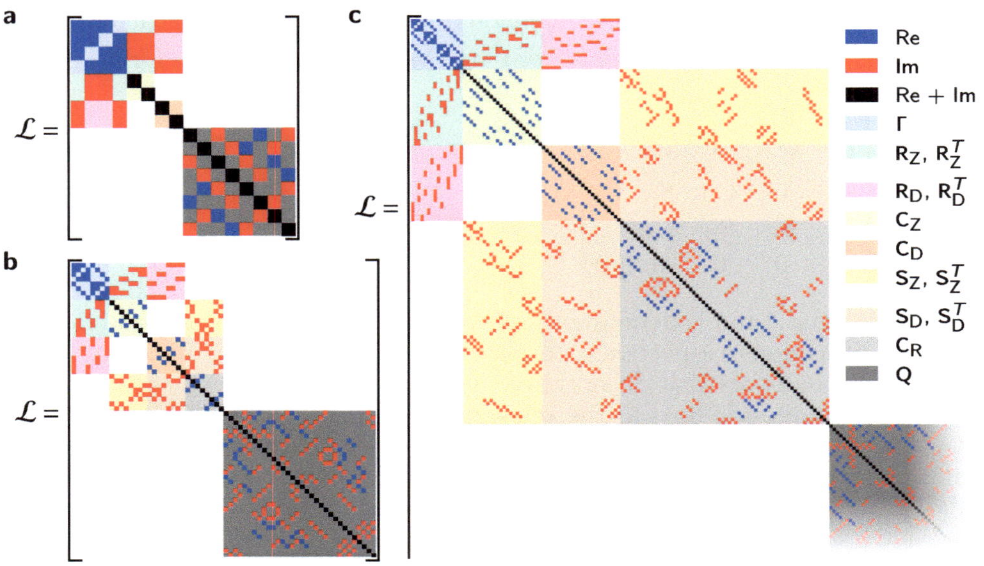

Figure 3.7: Liouvillian matrix structure. Liouvillian $\mathcal{L}$ in the block structure as stated in Eq. 3.53 for the qubit and **a** one TLS, **b** two, and **c** three TLSs. Non-zero real entries from $\Gamma^j_{\uparrow,\downarrow}$ are colored in blue, non-zero imaginary entries from the couplings g_k are colored in red, and non-zero real and imaginary entries from various relaxation, dephasing and frequency contributions are colored in black. Zero entries are colored according to the block structure. I would like acknowledge SymEngine and SymPy for providing a convenient computer algebra system. Reproduced and extended from Ref. [86].

and $\rho_{12} = \rho^*_{21}$. For the definition of the wave functions see Sec. 3.2 and App. B. In case the mutual decoherence $\Gamma_2 = \Gamma_2^{\mathrm{q}} + \Gamma_2^{\mathrm{t}}$ is large compared to the time scales on which the occupations ρ_{11} and ρ_{22} vary, one may approximate ρ_{21} by the population-dependent steady state value, which leads to overdamped vacuum Rabi oscillations (3.2). In view of Eq. 3.52, this means

$$\rho_{\mathrm{C}} = -\mathbf{C}^{-1}\mathbf{R} \cdot \rho_{\mathrm{D}}. \tag{3.55}$$

This approximation is best understood in the framework of Nakajima-Zwanzig's master equation [144, 148]. Solving the dynamics of the coherences in Eq. 3.52 gives

$$\rho_{\mathrm{C}}(t) = e^{\mathbf{C}(t-t_0)}\rho_{\mathrm{C}}(t_0) + \int_0^{t-t_0} e^{\mathbf{C}\tau}\mathbf{R}\,\rho_{\mathrm{D}}(t-\tau)\mathrm{d}\tau. \tag{3.56}$$

Insertion in the dynamics of the populations results in the integro-differential equation

$$\dot{\rho}_{\mathrm{D}}(t) = \mathbf{\Gamma}\rho_{\mathrm{D}}(t) + \mathbf{R}^T e^{\mathbf{C}(t-t_0)}\rho_{\mathrm{C}}(t_0) + \int_0^{t-t_0} \mathbf{R}^T e^{\mathbf{C}\tau}\,\mathbf{R}\,\rho_{\mathrm{D}}(t-\tau)\mathrm{d}\tau. \tag{3.57}$$

Under the assumption that the populations vary slowly compared to the relaxation of the coherences, several approximations can be performed. First of all, the second term can be neglected as we are only interested in the slow population dynamics on long time scales. It should be noted that it is often assumed instead that the coupling between system and environment is switched on at t_0. In this case, the initial system bath coherences vanish on average. However, in experiments the coupling to the environment is usually always on, which means that the initial coherences $\rho_C(t_0)$ will most likely not vanish in thermal equilibrium (Eq. 3.55). The only way to remove the system bath correlation is a projective measurement on the system. Next, we can also apply the first and second Markov approximations, as we did for the Bloch-Redfield equation (Sec. 3.3). This means to approximate ρ_D by its current time $\rho_D(t-\tau) \approx \rho_D(t)$ and to extend the integration to infinity, yielding

$$\begin{aligned}\dot{\rho}_D(t) &= \mathbf{\Gamma}\rho_D(t) + \mathbf{R}^T\mathbf{C}^{-1}e^{\mathbf{C}\tau}\mathbf{R}\Big|_0^\infty \rho_D(t) \\ &= \left(\mathbf{\Gamma} - \mathbf{R}^T\mathbf{C}^{-1}\mathbf{R}\right)\rho_D(t) = \mathcal{L}_D\,\rho_D(t). \end{aligned} \tag{3.58}$$

This proves that Eq. 3.55 follows from the two Markov approximations. For a quantitative discussion of the Markov approximations recall the derivation for the qubit and the TLS in Sec. 3.2.

If the qubit is coupled to more than one TLS, calculating the inverse of $\mathbf{C}$ is not feasible. However, since we assumed for the Markov approximation that the coherences vanish rapidly, the diagonal entries of $\mathbf{C}$ must be large in comparison to the off-diagonal entries. Therefore, the inverse can be approximated up to second order in the diagonal entries (App. D). If the TLSs are lossless (and not interacting with each other) the matrices $\mathbf{C}_{Z,D}$ are diagonal and the inverse can readily be computed. Otherwise, we need to approximate $\mathbf{C}_{Z,D}$ to be diagonal (App. D) to yield the Solomon equations. It holds

$$\mathbf{R}^T\mathbf{C}^{-1}\mathbf{R} \approx \mathbf{R}_Z^T\mathbf{C}_Z^{-1}\mathbf{R}_Z + \mathbf{R}_D^T\mathbf{C}_D^{-1}\mathbf{R}_D \tag{3.59}$$

$$\simeq \mathbf{R}_Z^T\mathbf{D}_Z^{-1}\mathbf{R}_Z + \mathbf{R}_D^T\mathbf{D}_D^{-1}\mathbf{R}_D. \tag{3.60}$$

Depending on the situation, one finds up to first or second order the cross-relaxation rates $\Gamma_{qt}^{\Delta_k}$ between the qubit and the kth TLS. The cross-relaxation rates are of the form [83, 84]

$$\Gamma_{qt}^{\Delta_k} = \frac{2g_k^2\Gamma_2^k}{(\Gamma_2^k)^2 + \Delta_k^2} \tag{3.61}$$

with $\Delta_k \in \{\delta_k, \sigma_k\}$. As a reminder, Γ_2^k is the sum of the qubit and TLS decoherence rates. The derivation of the formula can be tracked in Eq. 3.54 when the coherence ρ_{12} is approximated by its population-dependent steady state value. In the case of a qubit and a single TLS, where none of the previously discussed approximations are needed, the relaxation matrix can be written as

$$\mathcal{L}_{\mathrm{D}} = \begin{pmatrix} -\Gamma^{\mathrm{q}}_{\downarrow} - \Gamma^{\mathrm{t}}_{\downarrow} & \Gamma^{\mathrm{q}}_{\uparrow} & \Gamma^{\mathrm{t}}_{\uparrow} & 0 \\ \Gamma^{\mathrm{q}}_{\downarrow} & -\Gamma^{\mathrm{q}}_{\uparrow} - \Gamma^{\mathrm{t}}_{\downarrow} & 0 & \Gamma^{\mathrm{t}}_{\uparrow} \\ \Gamma^{\mathrm{t}}_{\downarrow} & 0 & -\Gamma^{\mathrm{q}}_{\downarrow} - \Gamma^{\mathrm{t}}_{\uparrow} & \Gamma^{\mathrm{q}}_{\uparrow} \\ 0 & \Gamma^{\mathrm{t}}_{\downarrow} & \Gamma^{\mathrm{q}}_{\downarrow} & -\Gamma^{\mathrm{q}}_{\uparrow} - \Gamma^{\mathrm{t}}_{\uparrow} \end{pmatrix} + \begin{pmatrix} -\Gamma^{\sigma}_{\mathrm{qt}} & & & \Gamma^{\sigma}_{\mathrm{qt}} \\ & -\Gamma^{\delta}_{\mathrm{qt}} & \Gamma^{\delta}_{\mathrm{qt}} & \\ & \Gamma^{\delta}_{\mathrm{qt}} & -\Gamma^{\delta}_{\mathrm{qt}} & \\ \Gamma^{\sigma}_{\mathrm{qt}} & & & -\Gamma^{\sigma}_{\mathrm{qt}} \end{pmatrix} \tag{3.62}$$

and similarly for more TLSs (s. App. E).

So far, we were able to reduce the 4^{n+1}-dimensional Liouvillian in Eq. 3.52 for the qubit and n TLSs to a 2^{n+1}-dimensional rate equation for the occupations. However, the goal is to reduce the rate equation even further to the $(n + 1)$-dimensional Solomon equations, describing the expectation values of the populations, i.e. the measurable population probabilities p_j. In general, one can not expect that there is an equation of motion for a closed set of expectation values, which is why several truncation strategies exist in the literature [149–153]. As will be shown in the next paragraph, the approximations in Eqs. 3.59 and 3.60 are sufficient to obtain the Solomon equations for an arbitrary number of TLSs, which is the central result of this derivation. For physical intuition, one can think of the approximation in Eq. 3.59 as neglecting coherence between TLSs, i.e. the TLSs are not cooperative, and in Eq. 3.60 as ignoring interference between different cross-relaxation pathways [146].

As already mentioned, the Solomon equations were originally derived for two nuclear spins, analogous to a qubit with a single TLS. In this case, the derivation is straightforward and was first presented by Solomon [83]. The general idea is to find a coordinate transformation $\mathbf{S}$ for ρ_{D} that leads to a covariant description of the probabilities. The probabilities are obtained by computing the corresponding partial traces. With our choice of ordering the diagonal entries (App. B), it holds that

$$p_{\mathrm{q}} = \sum_{\lfloor m/2 \rfloor \text{ integer}} \rho_{mm}, \qquad p^k_{\mathrm{t}} = \sum_{\lfloor m/2^k \rfloor \text{ integer}} \rho_{mm}, \qquad 1 = \sum_m \rho_{mm}. \tag{3.63}$$

Next, a basis transformation $\mathbf{S}$ is applied on ρ_{D} by defining $\rho'_{\mathrm{D}} = \mathbf{S} \cdot \rho_{\mathrm{D}}$ with $\rho'_{\mathrm{D}} = \begin{pmatrix} p_{\mathrm{q}} & p^1_{\mathrm{t}} & \dots & p^n_{\mathrm{t}} & 1 \,|\, \dots \end{pmatrix}^T$, leading to the dynamics $\dot{\rho}'_{\mathrm{D}} = \mathbf{S} \cdot \mathcal{L}_{\mathrm{D}} \cdot \mathbf{S}^{-1} \rho'_{\mathrm{D}}$. Note that the unity in the line above is required to describe the excitations from the ground state. As proven in App. E, on the basis of Eq. 3.60, the new rate equation is of the form

$$\mathbf{S} \cdot \mathcal{L}_{\mathrm{D}} \cdot \mathbf{S}^{-1} = \begin{pmatrix} \bar{\mathbf{A}} \quad \bar{\mathbf{\Gamma}}_{\uparrow} \quad \mathbf{0} \cdots \mathbf{0} \\ \hline \cdots \end{pmatrix}, \tag{3.64}$$

where $\bar{\mathbf{A}}$ is a $(n+1)$-dimensional square matrix and $\bar{\boldsymbol{\Gamma}}_\uparrow$ is a $(n+1)$-dimensional vector. They define the Solomon equations, which are independent of the microscopic structure of the populations, as reflected by the zeros to the right of $\bar{\boldsymbol{\Gamma}}_\uparrow$. Because of this decoupling, the matrix elements below the horizontal line are irrelevant for the dynamics of the probabilities. $\bar{\mathbf{A}}$ and $\bar{\boldsymbol{\Gamma}}_\uparrow$ describe the relaxation and excitation processes of the probabilities, respectively. The bar on top of $\bar{\mathbf{A}}$ and $\bar{\boldsymbol{\Gamma}}_\uparrow$ denotes that the rates are altered by the two-photon cross-relaxation rates $\Gamma_{\mathrm{qt}}^{\sigma_k}$. The Solomon equations comprise the following new relevant rates: the cross-relaxation rates $\bar{\Gamma}_{\mathrm{qt}}^{k} := \Gamma_{\mathrm{qt}}^{\delta_k} - \Gamma_{\mathrm{qt}}^{\sigma_k}$, the transition rates of the TLSs $\bar{\Gamma}_{\uparrow,\downarrow}^{\mathrm{t}_k} := \Gamma_{\uparrow,\downarrow}^{\mathrm{t}_k} + \Gamma_{\mathrm{qt}}^{\sigma_k}$, and the qubit transition rates $\bar{\Gamma}_{\uparrow,\downarrow}^{\mathrm{q}} := \Gamma_{\uparrow,\downarrow}^{\mathrm{q}} + \sum_k \Gamma_{\mathrm{qt}}^{\sigma_k}$. However, in the usual regime $\Gamma_2^k \ll \omega_{\mathrm{q}}$ for superconducting qubits, contributions from two-photon processes are negligible. In our scenario, where the TLSs are not interacting with each other, we obtain for $\bar{\mathbf{A}}$ a so-called arrow-head type matrix so that the Solomon equations are of the form

$$\dot{\mathbf{p}} = -\begin{pmatrix} \bar{\Gamma}_1^{\mathrm{q}} + \sum \bar{\Gamma}_{\mathrm{qt}}^{k} & -\bar{\Gamma}_{\mathrm{qt}}^{1} & \cdots & -\bar{\Gamma}_{\mathrm{qt}}^{k} \\ -\bar{\Gamma}_{\mathrm{qt}}^{1} & \bar{\Gamma}_1^{1} + \bar{\Gamma}_{\mathrm{qt}}^{1} & & \\ \vdots & & \ddots & \\ -\bar{\Gamma}_{\mathrm{qt}}^{n} & & & \bar{\Gamma}_1^{n} + \bar{\Gamma}_{\mathrm{qt}}^{n} \end{pmatrix} \mathbf{p} + \begin{pmatrix} \bar{\Gamma}_\uparrow^{\mathrm{q}} \\ \bar{\Gamma}_\uparrow^{1} \\ \vdots \\ \bar{\Gamma}_\uparrow^{n} \end{pmatrix} \tag{3.65}$$

with the usual definition $\bar{\Gamma}_1 = \bar{\Gamma}_\uparrow + \bar{\Gamma}_\downarrow$.

Finally, it should be noted that the proof in App. E is also valid when the TLSs are interacting with each other. This means that the Solomon equations are correct as long as the approximations that lead to the cross-relaxation rates in Eq. 3.61 can be justified. In general, it is very difficult to foresee the range of validity when the TLSs become coherent and close in frequency [154]. At some point a collective behavior of the TLSs comes into play, where the TLSs essentially form a large single spin, as discussed in detail in the context of superradiance [116, 117, 154], and as demonstrated in Sec. 3.6.2 by a stochastic Schrödinger equation simulation.

3.5 Analytical solutions and approximations

The rich relaxation dynamics of the Solomon equations allow us to draw conclusions on the connectivity and strength of the cross-relaxation rates. In NMR, the measured cross-relaxation rates contain information on the nuclear spins and their distances. This paved the way for the broad field of two-dimensional NMR spectroscopy [105]. Similarly, it will be shown in this section that the relaxation dynamics on long time scales contains information on the cross-relaxation distribution. From this distribution, one might be able to draw conclusions on the frequency and spatial distribution of the TLSs and get an idea of their physical nature. For instance, if there is a dipolar coupling between the qubit and the TLSs, one has $g_k \propto 1/r^3$, with r being the distance to the qubit. When the TLSs are spread on the surface around the qubit, we count the number $k(r) \propto r^2$ of TLSs within r and consequently $\Gamma_{\mathrm{qt}}^k \propto 1/r_k^6 \propto 1/k^3$. In case the TLSs are spread in all three dimensions one obtains $\Gamma_{\mathrm{qt}}^k \propto 1/k^2$. For the scenario presented in Fig. 3.6, when the coupling and the mutual decoherence are approximately constant but the TLSs are spread in frequency, the distribution $\Gamma_{\mathrm{qt}}^k \propto 1/k^2$ can be expected.

In order to simplify the discussion, it will be assumed that the mutual decoherence of the qubit and the TLSs is strong enough ($\Gamma_2^k > 4g_k$, s. Sec. 3.2) for the Solomon equations to be valid. For superconducting qubits this assumption is not always fulfilled. For instance, it is possible to observe coherent oscillations with dielectric TLSs [87, 155]. For weakly coupled TLSs, e.g. spins, this regime is applicable. When we compute the cross-relaxation distribution, it is further assumed that $\Gamma_2^k \ll \omega_{\mathrm{q}}$ such that two-photon processes $\Gamma_{\mathrm{qt}}^{\sigma}$ can be neglected (Eq. 3.61). Note that this assumption might not be valid for low-frequency fluxonium qubits [156]. Both assumptions and the form of the cross-relaxation rate Eq. 3.61 imply that the qubit will only exchange energy significantly with those TLSs that are close in frequency, $\delta_k \simeq \Gamma_2^k$. We will therefore assume that the qubit and the TLSs relax approximately to the same thermal population $p_{\mathrm{th}} = \Gamma_{\uparrow}^j/\Gamma_1^j$. Moreover, all TLSs are assumed to be of the same physical origin, suggesting a single intrinsic relaxation rate for the TLSs. Under these assumptions, we can now rewrite the Solomon equations: the bar on top of the rates is removed to indicate the neglect of two-photon processes, and for convenience the intrinsic relaxation rates $\Gamma_{\mathrm{q}} = \Gamma_1^{\mathrm{q}}$ and $\Gamma_{\mathrm{t}} = \Gamma_1^{\mathrm{t}_k}$ are introduced. The Solomon equations then read

$$\dot{\mathbf{p}} = -\begin{pmatrix} \Gamma_{\mathrm{q}} + \sum \Gamma_{\mathrm{qt}}^k & -\Gamma_{\mathrm{qt}}^1 & \cdots & -\Gamma_{\mathrm{qt}}^k \\ -\Gamma_{\mathrm{qt}}^1 & \Gamma_{\mathrm{t}}^1 + \Gamma_{\mathrm{qt}} & & \\ \vdots & & \ddots & \\ -\Gamma_{\mathrm{qt}}^n & & & \Gamma_{\mathrm{t}}^n + \Gamma_{\mathrm{qt}} \end{pmatrix} \mathbf{p} + \begin{pmatrix} \Gamma_{\mathrm{q}} \\ \Gamma_{\mathrm{t}} \\ \vdots \\ \Gamma_{\mathrm{t}} \end{pmatrix} p_{\mathrm{th}}. \tag{3.66}$$

Following from Eqs. 3.3 and 3.4, it holds for the qubit

$$\Gamma_{\uparrow}(t) = \Gamma_{\mathrm{q}} p_{\mathrm{th}} + \sum_k \Gamma_{\mathrm{qt}}^k p_{\mathrm{t}}^k(t), \tag{3.67}$$

$$\Gamma_{\downarrow}(t) = \Gamma_{\mathrm{q}}(1 - p_{\mathrm{th}}) + \sum_k \Gamma_{\mathrm{qt}}^k (1 - p_{\mathrm{t}}^k(t)), \tag{3.68}$$

$$\Gamma_1 = \Gamma_{\mathrm{q}} + \Gamma_{\mathrm{TLSs}}, \tag{3.69}$$

where it can be seen once again that the total qubit relaxation rate Γ_1 is independent of the TLS populations. The qubit relaxation induced by the TLSs is given by $\Gamma_{\mathrm{TLSs}} = \sum_k \Gamma_{\mathrm{qt}}^k$. From Eq. 3.5 it follows

$$p_{\mathrm{eq}}(t) = \frac{\Gamma_{\mathrm{q}} p_{\mathrm{th}} + \sum_k \Gamma_{\mathrm{qt}}^k p_{\mathrm{t}}^k(t)}{\Gamma_1}. \tag{3.70}$$

Similarly, one can introduce

$$p_{\mathrm{eq}}^{\mathrm{TLSs}}(t) = \frac{\sum_k \Gamma_{\mathrm{qt}}^k p_{\mathrm{t}}^k(t)}{\sum_k \Gamma_{\mathrm{qt}}^k}, \tag{3.71}$$

which would be the equilibrium population of the qubit in the absence of intrinsic qubit loss, and hence it serves as a measure for the TLS population.

The structure of the arrow-head matrix in the Solomon equations entails many useful properties [157, 158]. Without loss of generality, it can be assumed that the rates are non-zero $\Gamma_{\mathrm{qt}}^k > 0$ and sorted $\Gamma_{\mathrm{qt}}^k \geq \Gamma_{\mathrm{qt}}^{k+1}$. Using this definition, we can state one of the most important properties: The interlacing theorem yields immediately the relations

$$\lambda_0 > \Gamma_{\mathrm{qt}}^1 + \Gamma_{\mathrm{t}} \geq \lambda_1 \geq \cdots \geq \Gamma_{\mathrm{qt}}^n + \Gamma_{\mathrm{t}} \geq \lambda_n, \tag{3.72}$$

$$\lambda_0 > \Gamma_1 \geq \lambda_n \tag{3.73}$$

for the eigenvalues. Furthermore, $\lambda_n \geq \min\{\Gamma_{\mathrm{q}}, \Gamma_{\mathrm{t}}\}$ is equal if and only if $\Gamma_{\mathrm{q}} = \Gamma_{\mathrm{t}}$. The inequality (Eq. 3.72) shows that the cross-relaxation distribution translates approximately into the same distribution for the eigenvalues of the Solomon equations. In order to shorten the notations the out-of-equilibrium population $\mathbf{p}^* = \mathbf{p} - \mathbf{p}_{\mathrm{th}}$ is introduced in the following, where $\mathbf{p}_{\mathrm{th}} = \left(p_{\mathrm{th}}, \ldots p_{\mathrm{th}}\right)^T$ is the steady-state for the system in thermal equilibrium.

3.5.1 The case of identical cross-relaxation rates

The system of differential equations can be solved analytically for the special case when all cross-relaxation rates are identical, $\Gamma_{\mathrm{qt}}^k = \Gamma_{\mathrm{qt}}$. While this case is likely not relevant for dielectric TLSs in superconducting devices, it might be applicable for an environment consisting of hyperfine split spins given their narrow frequency distribution [61]. When the system is driven out of equilibrium by operating the qubit, e.g. following the polarization protocol described in Sec. 2.8, the TLSs will always be populated identically. This means that the $(n-1)$-fold eigenvalue $\lambda_1 = \Gamma_{\mathrm{t}} + \Gamma_{\mathrm{qt}}$ does not take part in the relaxation dynamics, which will therefore be bi-exponential. The analytical solutions are derived in App. F.

The relaxation of the qubit and the corresponding transition rates $\Gamma_\uparrow$ and $\Gamma_\downarrow$ are depicted in Fig. 3.8 for different numbers of TLSs with different initial populations. The qubit initialization, typically to the ground state $p_{\mathrm{q},0} = 0$ or excited state $p_{\mathrm{q},0} = 1$, leads to distinct long-term relaxation dynamics. This can be observed in particular for a small number of TLSs, as shown in Fig. 3.8a. The qubit initialization does not affect the starting value for the transition rates. However, since the initialization adds or removes one quanta in the system, we notice a difference in the relaxation on longer time scales, as shown in Fig. 3.8b. The bi-exponential relaxation in this

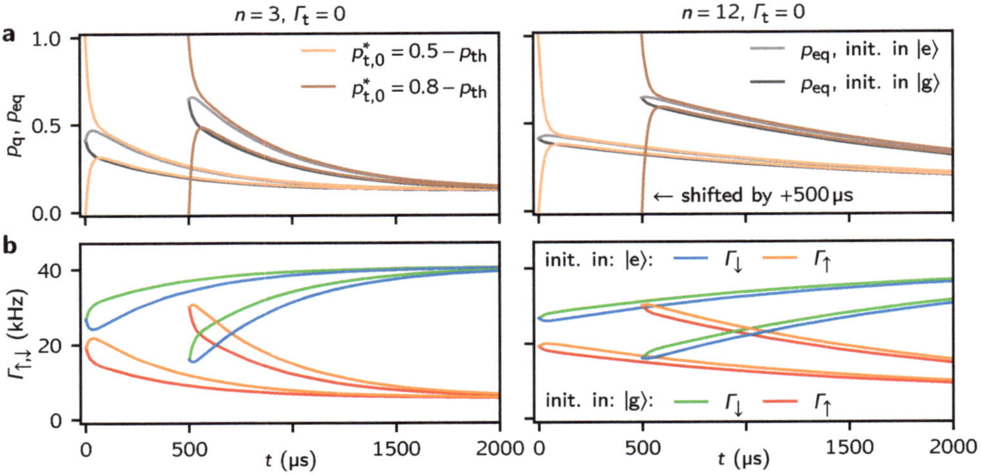

Figure 3.8: Qubit relaxation and transition rates for identical cross-relaxation rates. a Relaxation of the qubit population p_q and equilibrium population p_eq following qubit initialization to $|\mathrm{g}\rangle$ or $|\mathrm{e}\rangle$. All n TLSs are assumed to have the same initial population $p^*_{\mathrm{t},0}$. The curves with the higher initial TLS population are shifted in time for better visibility. Between the left and right panel, only the number n is varied. Additionally, a thermal population $p_\mathrm{th} = 0.12$ and lossless TLSs $\Gamma_\mathrm{t} = 0$ are assumed. The parameters were chosen to resemble the experimental findings [21]. **b** Qubit transition rates $\Gamma_{\uparrow,\downarrow}$ corresponding to the relaxation shown in panel a. Not surprisingly, the cooling and heating effect from the qubit initialization, i.e. the difference between e.g. the orange and red curves is less pronounced for increasing n. All curves are obtained from Eqs. F.3 and F.4, and using Eqs. 3.67, 3.68, and 3.70. Reproduced from Ref. [86].

scenario is qualitatively similar to the experimental results presented in Ref. [21], but insufficient to describe the slow and non-exponential relaxation observed on long time scales (cf. Fig. 3.10b). These effects can only be explained when the rates Γ^k_qt are not identical. Instead, a few fast and many slow cross-relaxation rates are required.

Considering the limit of infinitely many weakly coupled TLSs such that Γ_TLSs is constant, it holds $\lambda_{0,2} \in \{\Gamma_1, \Gamma_\mathrm{t}\}$. Then it follows from Eq. F.4 that the TLSs evolve independently of the qubit with $p^*_\mathrm{t}(t) = p^*_{\mathrm{t},0} e^{-\Gamma_\mathrm{t} t}$, while the qubit dynamics remains bi-exponential, governed by

$$\dot{p}_\mathrm{q} = -\Gamma_1 p_\mathrm{q} + \Gamma_\mathrm{TLSs} p^*_{\mathrm{t},0} e^{-\Gamma_\mathrm{t} t} + \Gamma_1 p_\mathrm{th}, \tag{3.74}$$

as may also read from the Solomon equations. To be more precise, in the interesting scenario in which the TLSs are long-lived, $\Gamma_\mathrm{t} < \Gamma_1$, we have $\lambda_2 = \Gamma_\mathrm{t} + \Gamma_\mathrm{TLSs}(\Gamma_\mathrm{q} - \Gamma_\mathrm{t})/(\Gamma_1 - \Gamma_\mathrm{t})/n + O(1/n^2)$, and for the TLSs it then holds that $p^*_\mathrm{t}(t) = p^*_{\mathrm{t},0} e^{-\lambda_2 t} + O(1/n)$. In the limit $n \gg 1$ with the TLSs being initially in thermal equilibrium $p^*_{\mathrm{t},0} = 0$, their relaxation thus becomes asymptotically decoupled from the qubit decay, justifying the Born approximation in this limit.

3.5.2 The case of distributed cross-relaxation rates

Continuing with the scenario of distributed cross-relaxation rates, it can be assumed without loss of generality that all the transfer rates are non-zero and distinct from each other. This is justified because if relaxation rates happen to be identical, their eigenvalues do not individually take part in the relaxation dynamics of the qubit and can be collapsed with Givens rotations, as was illustrated in the previous case analysis. The eigenvalues λ_m of an irreducible arrow-head matrix are given as the roots of the so-called Pick function [158]:

$$f(\lambda) = \Gamma_{\mathrm{q}} + \sum \Gamma_{\mathrm{qt}}^k - \lambda - \sum \frac{(\Gamma_{\mathrm{qt}}^k)^2}{\Gamma_{\mathrm{qt}}^k + \Gamma_{\mathrm{t}} - \lambda} = 0. \tag{3.75}$$

The corresponding eigenvectors can be expressed via

$$v_m = \begin{pmatrix} 1 & \Gamma_{\mathrm{qt}}^1/(\Gamma_{\mathrm{qt}}^1 + \Gamma_{\mathrm{t}} - \lambda_m) & \ldots & \Gamma_{\mathrm{qt}}^n/(\Gamma_{\mathrm{qt}}^n + \Gamma_{\mathrm{t}} - \lambda_m) \end{pmatrix}^T. \tag{3.76}$$

Fortunately, one can find that $\|v_m\|^2 = -\frac{\partial f}{\partial \lambda}\Big|_{\lambda_m}$. The fundamental solution of the Solomon equations from Sec. 3.5 can now be written as

$$\mathbf{p}(t) = \begin{pmatrix} 1/\|v_0\| & \ldots & 1/\|v_n\| \\ \vdots & & \vdots \\ & & \end{pmatrix} \cdot \begin{pmatrix} e^{-\lambda_0 t} & & \\ & \ddots & \\ & & \end{pmatrix} \cdot \begin{pmatrix} 1/\|v_0\| & \ldots & \\ \vdots & & \\ 1/\|v_n\| & \ldots & \end{pmatrix} \mathbf{p}^*(0) + \mathbf{p}_{\mathrm{th}}. \tag{3.77}$$

When initially only the qubit is out of equilibrium, it follows from Eq. 3.77

$$p_{\mathrm{q}}(t) = \sum_{m=0}^{n} \frac{e^{-\lambda_m t}}{\|v_m\|^2} p_{\mathrm{q},0}^* + p_{\mathrm{th}} \tag{3.78}$$

and from Eq. 3.70 with the help of Eq. 3.75 one finds

$$p_{\mathrm{eq}}(t) = \sum_{m=0}^{n} \frac{\Gamma_1 - \lambda_m}{\Gamma_1} \cdot \frac{e^{-\lambda_m t}}{\|v_m\|^2} p_{\mathrm{q},0}^* + p_{\mathrm{th}}, \tag{3.79}$$

where one may recognize Eq. 3.2 when inserting Eq. 3.78 in Eq. 3.79.

Polynomial relaxation

In the following, possible long-term relaxation behaviors of the system are discussed. Since p_{q} and p_{eq} decay in a similar way on long time scales (s. Eqs. 3.78 and 3.79), the qubit population $p_q(t)$ will be used to discuss the relaxation, which is directly accessible in experiments. In particular, we are interested in the emergence of non-exponential relaxation curves that must arise by virtue of Eq. 3.78 from the sum of many exponential functions. Obviously, if the qubit interacts with a finite number of TLSs, the relaxation can be approximated on long time scales by $p_q(t) \approx \alpha e^{-\lambda_n t}$, and the non-exponential behavior can only appear for $t < 1/\lambda_n$. An exponential relaxation will also be seen for an infinite number of TLSs that have a finite lifetime $\Gamma_t > 0$, in which case $p_q(t) \approx \alpha e^{-\Gamma_t t}$ is obtained on long time scales, as can be deduced from Eq. 3.72, and therefore, a non-exponential behavior can only appear for $t < 1/\Gamma_t$. Thus, a non-exponential relaxation appears on long time scales for a large number of TLSs that are long-lived, $\Gamma_t \approx 0$, and with eigenvalues λ_i that vanish continuously with $\lambda_i - \Gamma_t \to 0$.

As an insightful example, we will now discuss the experimental situation in which the TLSs are spread in frequency. For the modeling, the TLSs are assumed to be equally spaced in frequency with spacing Δ. In this case, the cross-relaxation rates (Eq. 3.61) are given by

$$\Gamma_{qt}^{h} = \frac{ab^2}{b^2 + (h - bc)^2} \quad \text{with} \quad h \in \mathbb{Z}, \tag{3.80}$$

where $a = 2g^2/\Gamma_2$, $b = \Gamma_2/\Delta$, $c = \Delta_0/\Gamma_2$, and Δ_0 describes a frequency shift of the TLSs with respect to the qubit frequency. Due to the periodicity, Δ_0 can be restricted to $\Delta_0 \in \{0, \Delta/2\}$. Here, the index h is used to clarify that these rates are not sorted (Eq. 3.72). In case $b \gg 1$, which corresponds to a high TLS density, many TLSs interact with the qubit with a similar rate. In this case, a good approximation on moderate time scales is given by the previous case study in Sec. 3.5.1. In the case $b \lesssim 1$, as in our experiment, we have the interesting situation in which the rates vanish continuously.

Next, the relaxation dynamics following from Eq. 3.78 is computed on long time scales. For the distribution in Eq. 3.80, the analysis below can only be applied for the special cases $\Delta_0 = 0$ (one TLS in resonance with the qubit), $\Delta_0 = \Delta/4$ (equally spaced detunings $|\delta_k|$), and $\Delta_0 = \Delta/2$ (maximum detuning between the closest two TLSs and the qubit). For more details see App. F.2. However, since we are mainly interested in the long time dynamics, we can slightly approximate the rate distribution to yield a simpler and more instructive analysis. We use

$$\Gamma_{qt}^{k} = a/k^2 \quad \text{with} \quad k \in \mathbb{N}^+ \tag{3.81}$$

$$\rightarrow \quad \Gamma_{\text{TLSs}} = \sum \Gamma_{qt}^{k} = \frac{a\pi^2}{6}. \tag{3.82}$$

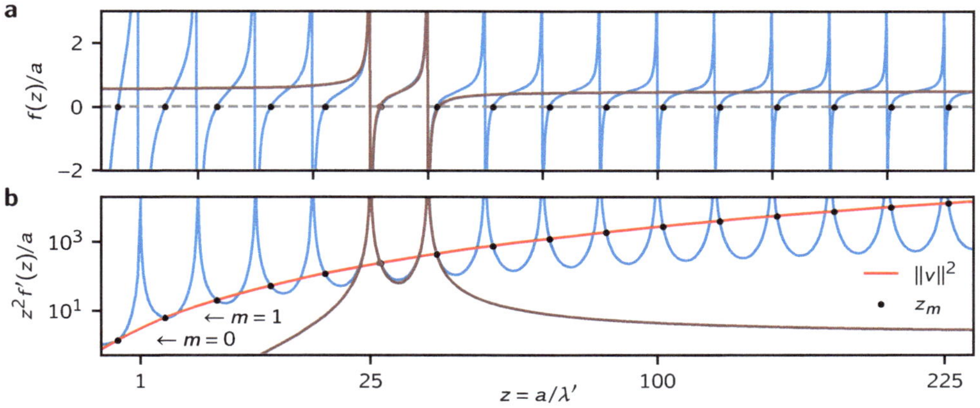

Figure 3.9: Pick function and its derivative. **a** Pick function $f(\lambda')/a$ and **b** its derivative $-f'(\lambda')/a$ plotted over the reciprocal coordinate $z = a/\lambda'$. Note the quadratic x-axis. The black points show the solution of the Pick function. The red curve shows the interpolation of the norms $\|v_m\|^2$ as given by Eq. 3.84. The local approximation of the Pick function with two poles is depicted by the brown curve. The parameters are chosen similar to the experiment with $a = 20\,\text{kHz}$, $\Gamma_q = 10\,\text{kHz}$, and $\Gamma_t = 0$. For more details see App. F.2.

Note that for a direct comparison with the distribution defined in Eq. 3.80, the parameter a needs to be scaled by a factor of 4. With the new distribution, the Pick function (Eq. 3.75) can be simplified and expressed in a closed form:

$$f(\lambda) = \Gamma_q - \Gamma_t - \frac{\lambda'}{2} - \frac{\pi a}{2}\sqrt{\frac{\lambda'}{a}}\cot\sqrt{\frac{\pi^2 a}{\lambda'}} \tag{3.83}$$

with $\lambda' = \lambda - \Gamma_t$. Furthermore, $\|v_m\|^2$ can be computed using the fact that by definition $f(\lambda_m) = 0$. We obtain

$$\begin{aligned} \|v_m\|^2 &= -f'(\lambda)\Big|_{\lambda_m} = -\frac{\partial f}{\partial \lambda'}\Big|_{\lambda'_m} \\ &= \frac{1}{2} + \frac{3\Gamma_{\text{TLSs}} - (\Gamma_q - \Gamma_t)}{2a}\frac{a}{\lambda'_m} + \frac{(\Gamma_q - \Gamma_t)^2}{a^2}\frac{a^2}{{\lambda'_m}^2} \\ &= 1/2 + \beta z_m + \gamma^2 z_m^2 . \end{aligned} \tag{3.84}$$

Here, we further introduced $z_m = a/\lambda'_m$. We will see that the linear and quadratic terms in z_m give rise to different long-term relaxation dynamics. The Pick function and its derivative are depicted in Fig. 3.9.

The next step is to evaluate the sum in Eq. 3.78 on long time scales. The derivation is presented in App. F. In the situation in which $\gamma^2 = 0$, which essentially describes the dilution of the initial qubit excitation into the TLS environment, the long-term relaxation dynamics is governed by

$$p^*_{\mathrm{q,approx.}}(t) = \lim_{t\to\infty} p^*_{\mathrm{q}}(t) = \frac{\sqrt{\pi}}{2\beta}\frac{e^{-\Gamma_{\mathrm{t}}t}}{(at)^{1/2}}\, p^*_{\mathrm{q},0}. \tag{3.85}$$

In the experimentally more likely situation in which $\gamma^2 > 0$, the relaxation dynamics becomes

$$p^*_{\mathrm{q,approx.}}(t) = \lim_{t\to\infty} p^*_{\mathrm{q}}(t) = \frac{\sqrt{\pi}}{4\gamma^2}\frac{e^{-\Gamma_{\mathrm{t}}t}}{(at)^{3/2}}\, p^*_{\mathrm{q},0}. \tag{3.86}$$

The long time relaxation dynamics is likely to be hidden in the noise when initially only the qubit is brought out of equilibrium. A simple way to improve the visibility is to also initialize a few of the most resonant TLSs, as in the experiment (s. Fig. 2.21). Then, for a large number of TLSs where we have $\lambda'_m \to 0$, one can deduce from Eqs. 3.76 and 3.77 that the long time solution is still valid. We just have to replace $p^*_{\mathrm{q},0}$ with the total out-of-equilibrium excitation $\sum_j p^*_{j,0}$. The long-term solution sets in under the condition $t \gg 1/\lambda_n$ with n such that $p^*_{k,0} = 0$ for $k > n$. In Fig. 3.10a several relaxation curves are shown verifying the power-law decay on long time scales, which can also be observed in the experiment (s. Fig. 3.10b). However, in the experiment the above condition is only approximately fulfilled, since all TLSs were at least partially excited by the polarization sequence (s. Fig. 2.21).

Generalization

As discussed at the beginning of this chapter, one can also expect other cross-relaxation distributions in experiments. In the following, we will therefore discuss the long time behavior for a general cross-relaxation distribution of the form

$$\Gamma^k_{\mathrm{qt}} = a/k^d \quad \text{with } d > 1 \text{ and } k \in \mathbb{N}^+ \tag{3.87}$$

$$\rightarrow \quad \Gamma_{\mathrm{TLSs}} = \sum \Gamma^k_{\mathrm{qt}} = \zeta(d) \tag{3.88}$$

with ζ being the Riemann zeta function. The derivation of the limit behavior is similar to the one presented above and is given in App. F. The difficulty is to determine the value of β, for which analytical expressions can be derived for integer values of d, as shown in App. F.2. For $\gamma = 0$, the relaxation dynamics approaches

$$p^*_{\mathrm{q,approx.}}(t) = \lim_{t\to\infty} p^*_{\mathrm{q}}(t) = \frac{\Gamma(1+\frac{1}{d})}{\beta}\frac{e^{-\Gamma_{\mathrm{t}}t}}{(at)^{1/d}}\, p^*_{\mathrm{q},0}. \tag{3.89}$$

Here, Γ denotes the Gamma function. For $\gamma^2 > 0$ it becomes

$$p^*_{\mathrm{q,approx.}}(t) = \lim_{t\to\infty} p^*_{\mathrm{q}}(t) = \frac{\Gamma(2-\frac{1}{d})}{\gamma^2 d}\frac{e^{-\Gamma_{\mathrm{t}}t}}{(at)^{2-1/d}}\, p^*_{\mathrm{q},0}. \tag{3.90}$$

For $d \gg 1$, this solution sets in very late and will be difficult to observe in the experiment unless the system is approximately lossless, i.e. $\gamma \to 0$.

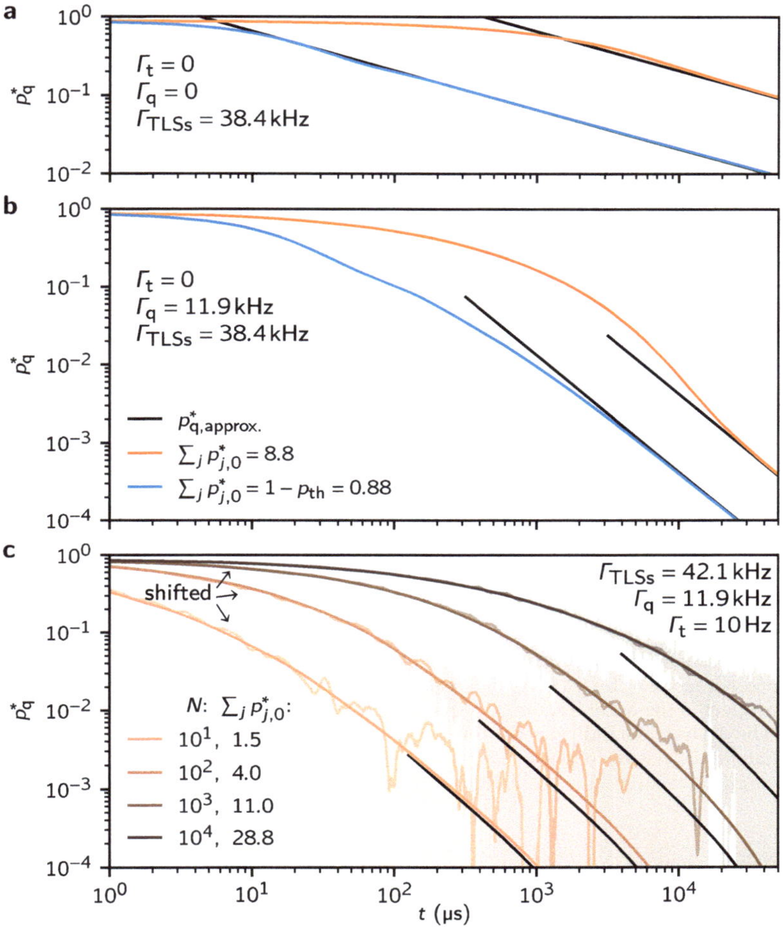

Figure 3.10: Polynomial relaxation. a, b Qubit relaxation with only the qubit being excited initially (blue lines) or with the first 9 TLSs being excited in addition (orange lines). The black curves show the limit solution of either Eqs. 3.85 or 3.86. We use the distribution $\Gamma^k_{qt} = 4ab^2/k^2$ with a and b as in panel c. Panel a shows the scenario without any losses in the system $\Gamma_q = \Gamma_t = 0$, simulated with 2000 TLSs, whereas panel b depicts the scenario with the intrinsic qubit loss Γ_q as in panel **c** and simulated with 100 TLSs. **c** Qubit relaxation taken from Sec. 2.8 for different lengths of the TLS polarization sequence to the excited state. To reduce the noise, a 5th-order Savitzky-Golay filter with increasing window length $w(t) = t/2$ was applied to the data. The curves $N = 10^1$ - 10^3 are shifted leftwards for better visibility by factors of $\sqrt{10}$. The continuous lines show the exact result (Eq. 3.77) using the distribution of Eq. 3.80 with $a = 25.5$ kHz, $b = 0.48$ and $c = 0$, slightly updated compared to Sec. 2.8. The initial polarization is modeled with Eq. 2.25 as shown in Fig. 2.21. The simulation was carried out with 101 TLSs, with more than 50 TLSs required for convergence. The limit solution Eq. 3.86 deviates with increasing initial polarization, since the condition $t \gg 1/\lambda_n$ with $p^*_{k,0} = 0$ for $k > n$ is not fulfilled (s. text and Fig. 2.21). Reproduced from Ref. [86].

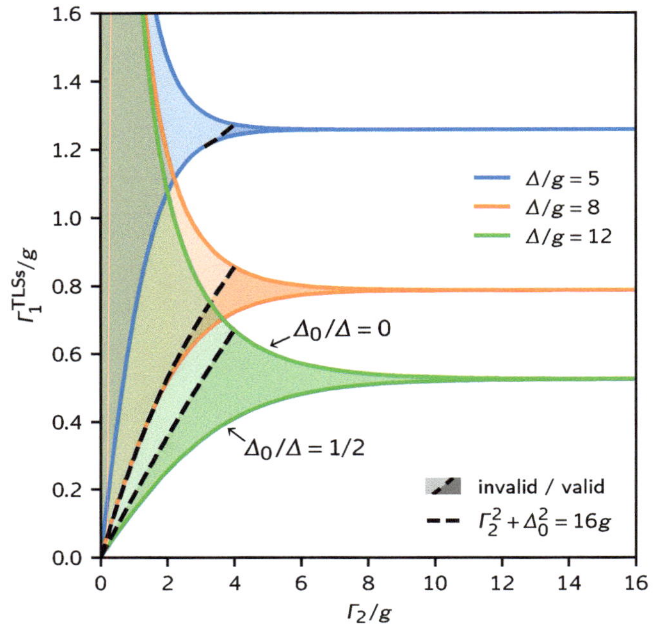

Figure 3.11: TLS induced qubit relaxation. Total cross-relaxation rate Γ_{TLSs} as a function of the mutual decoherence Γ_2 for various TLS frequency densities $1/\Delta$. The shift of the TLS ladder with respect to the qubit frequency is encoded in Δ_0, i.e. the detuning to the most resonant TLS. The dashed line indicates the threshold above which the calculation of the cross-relaxation rate Eq. 3.50 is no longer valid. In this region the qubit and the most resonant TLSs undergo damped vacuum Rabi oscillations (for more details see Sec. 3.2). Reproduced from Ref. [86].

3.5.3 Qubit relaxation as a function of the mutual decoherence

This section discusses the qubit relaxation as a function of the mutual decoherence Γ_2. It follows from Eq. 3.61 that the added qubit relaxation induced by a single TLS is a non-monotonic function in Γ_2, which vanishes in both limits of $\Gamma_2 \longrightarrow 0$ and $\Gamma_2 \longrightarrow \infty$. The contribution to the relaxation is maximized when $\Gamma_2 = \delta$, amounting to $\Gamma_{\mathrm{qt}} = g^2/\delta$. When the TLSs are spread in frequency with a single coupling strength g and mutual decoherence rate Γ_2, the sum of the distribution in Eq. 3.80 can be evaluated analytically and results in

$$\begin{aligned} \Gamma_{\mathrm{TLSs}} &= \sum_{h=-\infty}^{\infty} \frac{ab^2}{b^2 + (h - bc)^2} \\ &= \pi ab \, \frac{\sinh(2\pi b)}{\cosh(2\pi b) - \cos(2\pi bc)} \end{aligned} \tag{3.91}$$

with a, b and c as defined in Eq. 3.80. Since Γ_2 cancels in the prefactor $ab = 2g^2/\Delta$, the result of the Bloch-Redfield master equation can be recovered for sufficient decoherence such that $b \gtrsim 1$:

$$\Gamma_{\mathrm{TLSs}} \approx \frac{2\pi g^2}{\Delta} = 2\pi g^2 \nu_{\mathrm{t}}, \tag{3.92}$$

with ν_{t} being the TLS density. In Fig. 3.11 the qubit relaxation is plotted as a function of the mutual decoherence for various TLS densities. Note that if the qubit is mainly coupled to a single TLS, the qubit relaxation can be improved by increasing the decoherence of the TLS, e.g. by changing the ambient temperature or by applying saturation pulses to the TLS. However, this improvement is lost for multiple TLSs, since the spectral broadening due to the increased Γ_2 exposes the qubit to far detuned TLSs. Another method for increasing Γ_2 and changing qubit relaxation is to exploit the photon shot noise dephasing during qubit readout, as recently demonstrated in Ref. [159].

3.6 Non-Poissonian quantum jumps

Understanding the quantum jump statistics of a qubit can give valuable insight into the real-time dynamics, which is lost when looking at the statistical density matrix. For instance, when the qubit is in contact with a Born-Markovian environment, the quantum jumps are Poisson-distributed, i.e. the qubit undergoes an exponential relaxation. When the qubit is coupled to a finite-size TLS environment, a non-Poissonian statistics can be expected due to measurement-induced temperature fluctuations in the TLS environment (Fig. 3.14). The non-Poissonian statistics is visualized by taking a histogram of the qubit dwell times in the excited or ground state (Fig. 3.12). The quantum jump statistics are extracted from quantum jump traces obtained by stroboscopic projective qubit measurements at equidistant intervals, as introduced in Sec. 2.7.2. For the extraction, the stroboscopic quantum jump traces that have been recorded for the qubit relaxation measurements can be used, however only taking the part where the qubit is in equilibrium. In this way, the extracted system parameters can be used to model the quantum jump statistics.

In this section, it will be shown that the measured quantum jump statistics can be reproduced using a diffusive stochastic Schrödinger equation (SSE) (Fig. 3.17) and also, somewhat surprisingly, using the Solomon equations (Fig. 3.18). While the SSE offers a more complete picture by tracking the entanglement of the qubit and the TLS system, the stochastic evolution of the wave function becomes computationally demanding with increasing system size. In addition, the SSE is not unique as it depends on the system details. In a nutshell, the quantum mechanical challenge is to describe the flow of excitation and energy during the measurement process, in particular when additional environments of the qubit [7] and the TLSs are included.

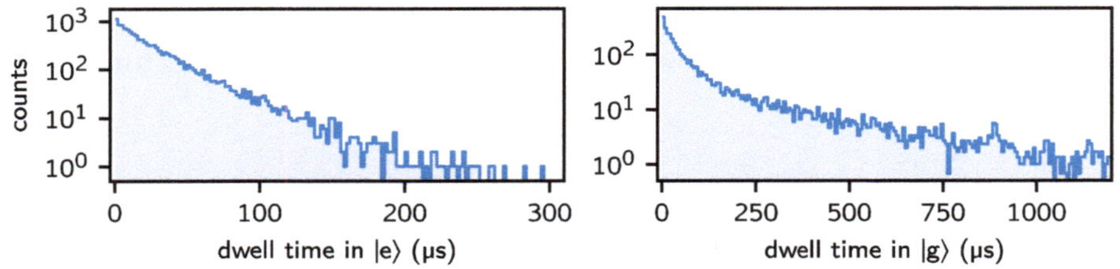

Figure 3.12: Non-Poissonian quantum jumps. Measured non-Poissonian quantum jump statistics of the fluxonium qubit dwell times in the excited state ($|e\rangle$, left panel) and the ground state ($|g\rangle$, right panel). The quantum jump statistics is extracted from 100 quantum jump traces of the $N = 1$ and initialization to $|g\rangle$ experiment and using the last 40 ms where the qubit is in thermal equilibrium. The total number of counts was rounded off to 12000. The bin size corresponds to the measurement repetition time of $t_{\mathrm{rep}} = 2\,\mu\mathrm{s}$. For $|g\rangle$, the histogram was binned by a factor of three in order to reduce the noise. Fluctuations of the statistics are discussed in Sec. 3.6.6 and are given as a video in Ref. [160]. Reproduced from Ref. [86].

Using the Solomon equations implicitly assumes that the measurement back action on the TLSs does not change their population expectation values. Therefore, one can not expect an agreement with the SSE in general. Indeed, as will be shown numerically in Sec. 3.6.4 that both procedures can lead to distinct quantum jump distributions for a small number of TLSs that dominate intrinsic qubit loss. This allows us to differentiate between quantum and classical behavior. Measuring the quantum jump statistics can therefore be used to identify a reduced measurement back action as well as to investigate a quantum-to-classical transition with an increasing number of TLSs.

Before proceeding with the experimental results, the role of the measurement back action will be clarified in the following. At this point, it should be emphasized that the TLSs in the experiment, while being approximately lossless, provide the main loss mechanism for the qubit. For now, additional environments are therefore neglected and the closed qubit-TLS system is considered. Let $|\psi_{\mathrm{i}}\rangle$ and $|\psi_{\mathrm{f}}\rangle$ denote the wave functions before and after a projective qubit measurement, respectively. Following a textbook quantum mechanical measurement on the qubit, $|\psi_{\mathrm{f}}\rangle$ will be a product state with the qubit being projected either to $|g\rangle$ or $|e\rangle$, and hence

$$|\psi_{\mathrm{f}}\rangle = \begin{cases} |e\rangle \otimes |\psi'_{\mathrm{TLSs}}\rangle & \text{with prob. } p_{\mathrm{q}}, \\ |g\rangle \otimes |\psi''_{\mathrm{TLSs}}\rangle & \text{with prob. } 1 - p_{\mathrm{q}}, \end{cases} \tag{3.93}$$

with the TLS wave function altered depending on the measurement outcome. In this setting, three scenarios representing very different back action properties will be highlighted in the following.

The first scenario shows that the excitation difference of the qubit before and after the measurement can be provided by the TLSs. Let $\mathcal{M}(m)$ denote the mth excitation manifold. Then, if $|\psi_{\mathrm{i}}\rangle \in \mathcal{M}(m)$, it follows that $|\psi_{\mathrm{f}}\rangle \in \mathcal{M}(m)$ and the excitation number is conserved. Here, the measurement process does not change the excitation number, but if the qubit and TLS photon energies are different, it must account for the energy difference.

The second scenario is essentially opposite to the first and contains those wave functions where the expectation values of the TLS populations remain unchanged by the qubit measurement. For example, suppose that the qubit and the TLSs are in the non-entangled product state, then it holds

$$|\psi_\mathrm{i}\rangle = \left(\sqrt{1-p_\mathrm{q}}\,|\mathrm{g}\rangle + \sqrt{p_\mathrm{q}}e^{i\varphi}\,|\mathrm{e}\rangle\right) \otimes |\psi_\mathrm{TLSs}\rangle \tag{3.94}$$

$$\rightarrow \quad |\psi_\mathrm{f}\rangle = \begin{cases} |\mathrm{e}\rangle \otimes |\psi_\mathrm{TLSs}\rangle & \text{with prob. } p_\mathrm{q}, \\ |\mathrm{g}\rangle \otimes |\psi_\mathrm{TLSs}\rangle & \text{with prob. } 1-p_\mathrm{q}. \end{cases} \tag{3.95}$$

Here, the excitation difference of the qubit must be provided entirely by the measurement process.

Lastly, in general, in the third scenario, the excitations can change arbitrarily. Consider, for instance, the following wave function, which is a mixture of two excitation manifolds:

$$|\psi_\mathrm{i}\rangle = \sqrt{1-p_\mathrm{q}}\,|\mathrm{g}\rangle \otimes |\varphi\rangle + \sqrt{p_\mathrm{q}}e^{i\varphi}\,|\mathrm{e}\rangle \otimes |\chi\rangle\,, \tag{3.96}$$

where $|\varphi\rangle \in \mathcal{M}(m)$ and $|\chi\rangle \in \mathcal{M}(n)$. After the measurement, the final wave function will be either in $|\psi_\mathrm{f}\rangle \in \mathcal{M}(m)$ or in $|\psi_\mathrm{f}\rangle \in \mathcal{M}(n+1)$. Here, we see that the total excitation difference must be provided by the measurement process, and that repeated measurements on the qubit have a tendency to steer the system into an excitation manifold. For several TLSs in higher excitation manifolds, this purification will become less effective and will likely not compete with relaxation processes that provide transitions between neighboring manifolds. For an experimental verification of a spin bath purification, see Ref. [98].

3.6.1 Stochastic Schrödinger equation

The following sections outline the simulations with the stochastic Schrödinger equation (SSE). These simulations clarify the previous discussion about the measurement back action, give insights into the polarization process, and can reproduce the measured non-Poissonian quantum jump statistics. The stochastic simulation is cross-checked with the Solomon equations, in which context a limit case of the Solomon equations is investigated as well.

Quantum trajectories must, on average, reproduce the dynamics of their underlying master equation. However, the unraveling of master equations is not unique [161], but can sometimes be motivated by the underlying physical processes [9, 122, 162]. A successful modeling of the measured quantum jump statistics may therefore distinguish different unravelings and could reveal a deeper understanding of the qubit environmental interaction.

Following the discussion in the previous section, we will first consider a closed system consisting of the qubit and the finite TLS environment under the presence of pure dephasing. A natural choice of modeling the dephasing is to think of it as a diffusive process as the result of a classical noise source, as explained in Sec. 3.2. In order to speed up the simulation, the random phase jump events in time can be combined to yield a stochastic Gaussian increment for each integration time step. In other words,

$$\int_0^{\mathrm{d}t} \eta(t')\mathrm{d}t' = \sqrt{\frac{\Gamma_\varphi}{2}}\mathrm{d}W, \tag{3.97}$$

where $\mathrm{d}W$ is the Wiener increment with variance $\mathrm{d}t$. To further speed up the simulation, the rotating-wave approximation can be applied for the $\sigma_x^{\mathrm{q}}\sigma_x^{\mathrm{k}}$ qubit TLS interaction (Eq. 3.1). This allows us to keep the integration time step in the interaction picture on an acceptable level, in the order of the coupling strength and decoherence. Due to the longitudinal noise, transformation to the interaction picture can be performed exactly via

$$|\psi_{\mathrm{I}}(t)\rangle = \prod_j e^{i\left(\frac{\omega_j t}{2} + \sqrt{\frac{\Gamma_\varphi^j}{2}} W_t^j\right)\sigma_z^j} |\psi(t)\rangle\,, \tag{3.98}$$

with W_t^j denoting independent Wiener processes for the qubit and the TLSs.

3.6.2 Limits of the Solomon equations by cooperative TLSs

In Fig. 3.13 various SSE simulations of the qubit relaxation are shown and compared to the predictions of the Solomon equations. Of particular interest is the validity of the Solomon equation with increasing coherence in the TLS reservoir. Since the SSE simulation presented so far is lossless, and due to the rotating-wave approximation, the relaxation occurs independently in all excitation manifolds. To fill up all these manifolds the system is initialized in a product state with populations $p_{\mathrm{q}} = 0.9$ and $p_{\mathrm{t}}^k = 0.1$ and random phases. The Solomon equations can still give accurate results even in the limit of infinite TLS coherence, provided that the TLSs are spectrally broadened. As a typically rule of thumb, deviations can be expected on time scales on the order of the inverse TLS energy splitting [134]. This estimation can probably be made more precise by regarding also the coupling strength. In the limit of vanishing spectral broadening, the TLSs behave cooperatively and form a single spin with larger quantum number. As a consequence, the qubit can only exchange energy with a single photon subspace of the Hilbert space and decays to only ~ 50 % on long

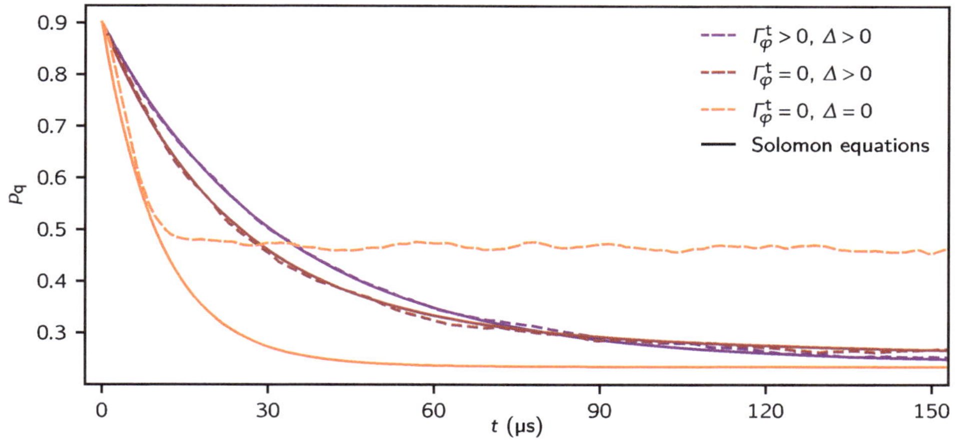

Figure 3.13: SSE simulation of the qubit relaxation with increasing TLS coherence. SSE simulation (dashed curves) and prediction by the Solomon equations (solid curves) of the qubit relaxation for a reservoir consisting of 5 TLSs symmetrically spaced in frequency around the qubit frequency. The parameters have been chosen similar to the experiment with $\Gamma_\varphi^q = 0.5\,\mathrm{MHz}$, $g = 2\pi \cdot 10\,\mathrm{kHz}$ and $\Delta = 2\pi \cdot 100\,\mathrm{kHz}$. For the incoherent scenario (purple curves) a TLS dephasing of $\Gamma_\varphi^{t_k} = \Gamma_\varphi^q$ is used, while for the more coherent scenarios the TLS dephasing is set to zero. For the spectrally coherent TLS environment (orange curves) the TLS energy spacing Δ is set to zero in addition. The integration was performed with time steps $\mathrm{d}t = 100\,\mathrm{ns}$ and the relaxation is based on 500 trajectories.

time scales. In this regime, the derivation of the Solomon equations is no longer valid, since the inverse of $\mathbf{C}$ can no longer be computed as some of the diagonal entries of $\mathbf{C}_\mathrm{R}$ vanish (Sec. 3.4).

3.6.3 Quantum jumps of a lossless system

A stroboscopic quantum jump trace simulated with the SSE is shown in Fig. 3.14. To illustrate the different back action scenarios discussed in the beginning of Sec. 3.6, the system wave function was initialized in a product state with populations $p_j = 0.2$ and random phases. The fluctuating and varying excitation in the system comes solely from the measurement process indicating mainly scenario three. The second scenario is only realized for the first measurement. Eventually, these fluctuations come to an end when the system state is trapped in one of the photon manifolds; in the case shown here, the system is trapped in the two-photon manifold. From here on, the dynamics is described by the first scenario.

Next, the SSE is used to visualize the polarization process (Fig. 3.15). For this, the measurement is followed by an instantaneous reset of the qubit to its target state. It turns out that, on average, the polarization follows the prediction of the Solomon

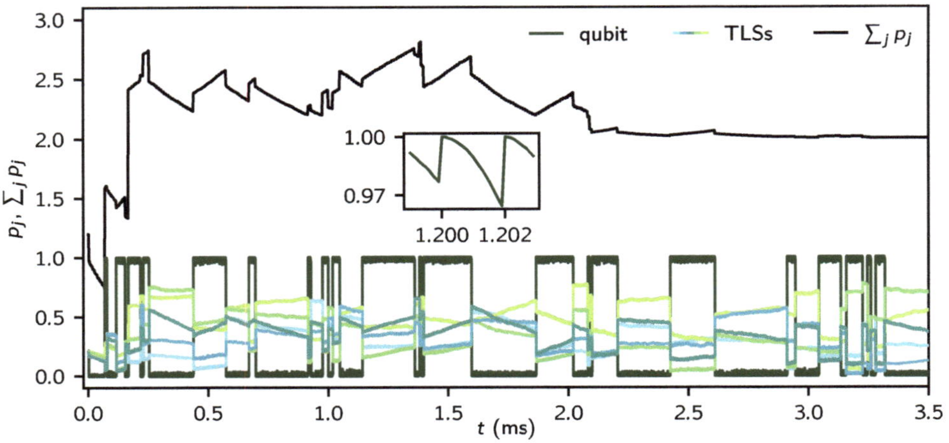

Figure 3.14: Illustration of the measurement back action. Simulation of a stroboscopic quantum jump trace using the stochastic Schrödinger equation. The qubit is measured with a repetition time $t_{\text{rep}} = 2\,\mu\text{s}$, as can be seen in the inset. At $t = 0$, the system is in a product state with populations $p_j = 0.2$ and random phases. Eventually, the system will be trapped, here in the two-photon manifold, where the total population (black lines) remains constant. The parameters are the same as of the incoherent scenario in Fig. 3.13, except that a lower TLS dephasing $\Gamma_\varphi^{t_k} = 0.1\,\text{MHz}$ is used here. Adapted from Ref. [86].

equations (Eq. 2.25), for which there is probably a simple explanation that considers the measurement back action. However, since for moderate polarization times the fluctuations in the polarization translate to correlated TLS populations, a precise modeling of the qubit relaxation following this polarization process must ideally also include these correlations.

3.6.4 Quantum jumps of a lossy system

Due to the trapping behavior (Fig. 3.14), the lossless SSE is not suited to simulate longer quantum jump traces with a specific average qubit population corresponding to a certain temperature. In reality, this temperature is defined by additional qubit environments that are responsible for its intrinsic relaxation, essentially providing transitions between excitation manifolds of the qubit-TLS system. Since the simulation is already computationally very expensive for a few TLSs, including additional degrees of freedom seems hopeless.

We must therefore find a stochastic simulation scheme that can emulate the intrinsic qubit environment. Ideally, it should also incorporate the far detuned long-lived TLSs that are seen in the experiment but cannot be included in the simulation for the very same reason. Thinking of this mind-blowingly large Hilbert space consisting of all the TLSs and also the intrinsic qubit environment and with all these photon

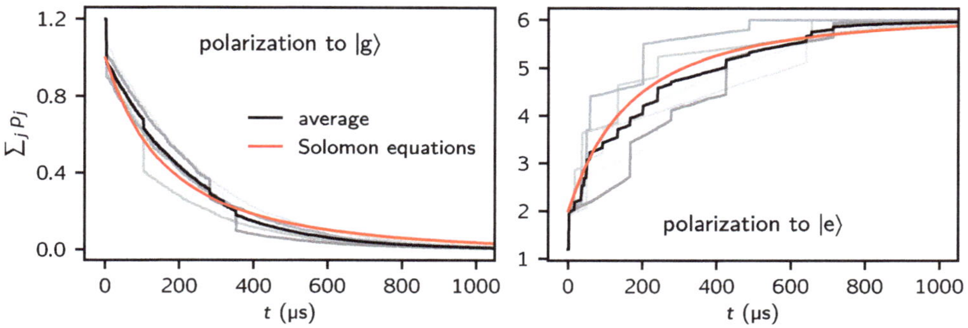

Figure 3.15: Illustration of the polarization process. Various SSE simulated trajectories of the total system excitation $\sum_j p_j$ during qubit polarization to $|g\rangle$ and $|e\rangle$ in the left and right panel, respectively. On average, the trajectories correspond to the prediction of the Solomon equations (Eq. 2.25). The parameters for the simulation are the same as used in Fig. 3.14.

number conserving interactions (in the rotating-wave approximation), it becomes clear that transitions between excitation manifolds, in case they are generated by the measurement, will not change the dynamics fundamentally. This means that the non-Poissonian quantum jump statistics must appear already within an excitation manifold. Confining the dynamics to one excitation manifold allows us to consider more elements, just enough to include the most resonant TLSs and to model the intrinsic qubit environment.

These reduced simulations are then compared with the second approach, a diffusive SSE that includes qubit losses. However, since both stochastic simulations capture the measurement back action, they still scale exponentially with the number of elements. For this reason, a third, much faster approach is presented in the next section. It involves simulating quantum jumps using the Solomon equations, which neglects the measurement back action in the system.

Beginning with the first approach, we decided to investigate a system consisting of 16 elements trapped in the three-photon manifold. Since all states will be equally populated on average, a finite temperature can be modeled corresponding to a thermal population of $p_{\text{th}} = 3/16$, similar to the experiment. To illustrate that the SSE and the Solomon equations lead to distinct non-Poissonian quantum jump distributions, it is considered here that the entire measured cross-relaxation Γ_{TLSs} originates from only one or two long-lived TLSs in resonance with the qubit. The remaining 14 or 13 weakly coupled TLSs, which can for instance be imagined as dielectric TLSs, are used to emulate the measured intrinsic qubit loss. They are also assumed to be on resonance with the qubit. For all TLSs, a dephasing rate $\Gamma_{\varphi}^{t_k} = 1.0\,\text{MHz}$ is used, which removes a minor coherent onset in the quantum jump distribution that can be seen when the TLS coherence exceeds the measurement repetition time (s. Supp. in Ref. [86]). For this reason, the integration time step was

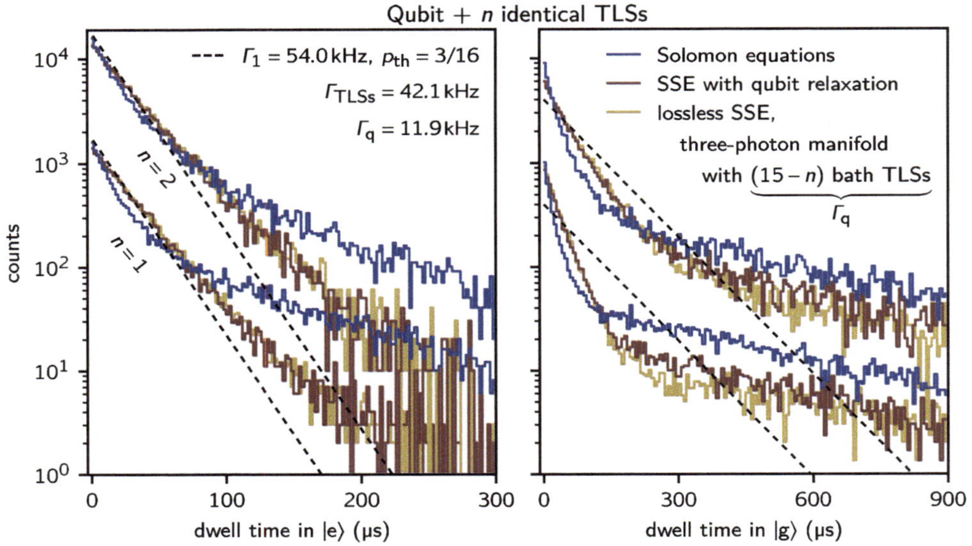

Figure 3.16: Comparison of various simulated quantum jump statistics. a Simulated quantum jump statistics of the qubit dwell times in $|e\rangle$ (left panel) and $|g\rangle$ (right panel), respectively, assuming that the qubit is coupled strongly to $n = 1$ or 2 TLSs that are in resonance with the qubit such that their cross-relaxation adds up to the total measured cross-relaxation Γ_{TLSs}. The distributions for $n = 2$ are shifted upwards by a factor of 10 for better visibility. The black dashed line shows the exponential distribution that would be obtained for a Born-Markovian qubit environment. In the lossless SSE simulation, the qubit couples in addition weakly to $(15 - n)$ TLSs in resonance with the qubit. They emulate the measured intrinsic qubit relaxation Γ_{q}. The simulation is carried out in the three-photon manifold. The thermal population was therefore set to $p_{\text{th}} = 3/16$ in all simulations. For the dephasing, the rates $\Gamma_\varphi^{\text{q}} = 0.5\,\text{MHz}$ and $\Gamma_\varphi^{\text{t}} = 1.0\,\text{MHz}$ are used. The stochastic simulations are carried out with $\mathrm{d}t = 1/32\,\mu\text{s}$. The histograms are generated as described in Fig. 3.12, except that all distributions are based on 20000 counts here. Reproduced from Ref. [86].

reduced to $\mathrm{d}t = 1/32\,\mu\text{s}$. The simulation in the three-photon manifold provides enough excitations to populate the qubit and the strongly coupled TLSs. The qubit will therefore experience a fluctuating temperature (s. Eq. 3.71) resulting from the random walk on the hypersphere. The simulation results are depicted in Fig. 3.16.

In the second approach, transitions between the excitation manifolds are induced by modeling the qubit relaxation with jump operators. They can be interpreted as measurements on the qubit's intrinsic environments. In the spirit of the previous lossless simulation, where the excitation diffuses between the qubit-TLS system and the intrinsic qubit environment, a diffusive SSE is needed where the energy diffuses between excitation manifolds. In contrast, in the commonly used quantum jump method, relaxation happens abrupt as a jump, which can be thought of as a photon detection [163]. The diffusive unraveling of the Lindblad equation was first introduced by Gisin and Percival [164], however, not in its most general form [161].

For a comprehensive derivation that shows the connection to the quantum jump method, see Ref. [165]. The state evolution can be computed as follows:

$$|\psi_\mathrm{I}(t+\mathrm{d}t)\rangle = |\psi_\mathrm{I}(t)\rangle - \frac{i}{\hbar}H_\mathrm{int}(t)\,|\psi_\mathrm{I}(t)\rangle\,\mathrm{d}t \tag{3.99}$$

$$+\sum_\alpha\left(\langle L_\alpha^\dagger\rangle L_\alpha - \frac{1}{2}L_\alpha^\dagger L_\alpha - \frac{1}{2}\langle L_\alpha^\dagger\rangle\langle L_\alpha\rangle\right)|\psi_\mathrm{I}(t)\rangle\,\mathrm{d}t \tag{3.100}$$

$$+\sum_\alpha\left(L_\alpha - \langle L_\alpha\rangle\right)|\psi_\mathrm{I}(t)\rangle\,\frac{\mathrm{d}W_\alpha}{\sqrt{2}}, \tag{3.101}$$

with $\langle L_\alpha\rangle = \langle\psi_\mathrm{I}|L_\alpha|\psi_\mathrm{I}\rangle$, the qubit jump operators $L_1 = \sqrt{\Gamma_\downarrow^\mathrm{q}}\cdot\sigma_\mathrm{q}^-$ and $L_2 = \sqrt{\Gamma_\uparrow^\mathrm{q}}\cdot\sigma_\mathrm{q}^+$, and independent complex Wiener increments $\mathrm{d}W_\alpha$. Each step is followed by a normalization of the wave function. The great advantage of this unraveling is its invariance when transforming between the Schrödinger and the interaction picture [161]. This allows the modeling of the dephasing as introduced in the beginning of Sec. 3.6.1. The complex noise makes perfectly sense when thinking of all these dephasing TLSs that add and extract bits of energy with random phases. In fact, the diffusive unraveling can also be interpreted to originate from a heterodyne photodetection of the system [122, 162]. Nevertheless, it is still surprising and satisfying to see (Fig. 3.16) that the simulation with the lossy SSE yields the same statistics than the lossless SSE in the enlarged Hilbert space and confined in one excitation manifold.

As already mentioned and and as will be explained in detail in the next section, the Solomon equations can also be used to simulate quantum jump traces. However, as can be seen in Fig. 3.16, the resulting statistics is in general distinct from each other. This shows that the measurement back action has a measurable effect on the real time quantum jump dynamics, which is lost when looking at the average relaxation dynamics. Strikingly, the measurement back action can be investigated without direct excess to the long-lived TLS environment. However, with increasing number of TLSs, the effect of the measurement back action vanishes, and the statistics becomes exponential.

3.6.5 Comparison to experiment

After having verified that the lossy SSE can emulate the intrinsic qubit environment in form of weakly coupled dephasing TLSs, we now turn to the experimental situation. Fig. 3.17 shows the results of the lossy SSE simulation, which reproduces the measured quantum jump statistics. The computational complexity could be kept on an acceptable level by recognizing that in particular the most resonant TLS contributes dominantly to the non-Poissonian quantum jump statistics. Accordingly, the simulation was performed for the qubit and the seven most resonant TLSs.

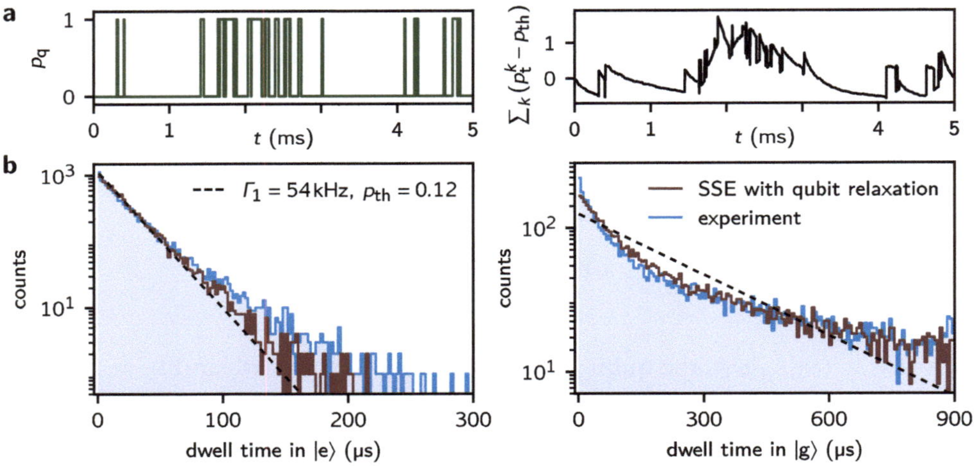

Figure 3.17: Comparison between SSE-simulated and measured quantum jumps. **a** SSE simulation of a quantum jump trace similar to Fig. 3.14, but including qubit relaxation and the seven most resonant TLSs. The cross-relaxation of the remaining TLSs was added to the intrinsic qubit loss. For the dephasing, the rates $\Gamma_\varphi^{q} = 0.5\,\mathrm{MHz}$ and $\Gamma_\varphi^{t} = 1.0\,\mathrm{MHz}$ are assumed. **b** Measured and SSE-simulated quantum jump statistics of the qubit dwell times in $|e\rangle$ and $|g\rangle$, respectively. The black dashed line shows the expected exponential distribution for a Born-Markovian qubit environment with qubit relaxation $\Gamma_1 = \Gamma_q + \Gamma_{TLSs}$ in thermal equilibrium corresponding to p_{th}. The histograms are generated as described in Fig. 3.12. Reproduced from Ref. [86].

In fact, a simulation with only the most resonant TLS and where all other TLSs are incorporated into the intrinsic qubit environment gives already a very good agreement with the experimental findings. This can be surmised from Fig. 3.16, which shows already similar distributions in comparison to the experiment. In addition, the resemblance of the distributions with those generated with the Solomon equations (Fig. 3.16) tells us that the effect of the measurement back action can in good approximation be neglected in the experiment, which brings us to the third approach, the modeling of the quantum jumps using the Solomon equations.

Suppose the measurement process is not changing the TLS populations, which means that the second back action scenario is realized. Then, the Solomon equations can be integrated to obtain the qubit population at the time of the upcoming measurement and yield the probabilities of the measurement outcome. Starting in thermal equilibrium, we can generate a quantum jump trace and compute the fluctuating energy in the TLS bath that in turn influences the measurement outcome of the qubit (s. Fig. 3.18a, b). Note the continuous TLS population in contrast to the SSE simulation in Fig. 3.17a that includes the measurement back action. The resulting martingales for the energies in the system are similar to Pólya's urn model, except that the qubit and the TLSs can (i) saturate and (ii) decay into their environments.

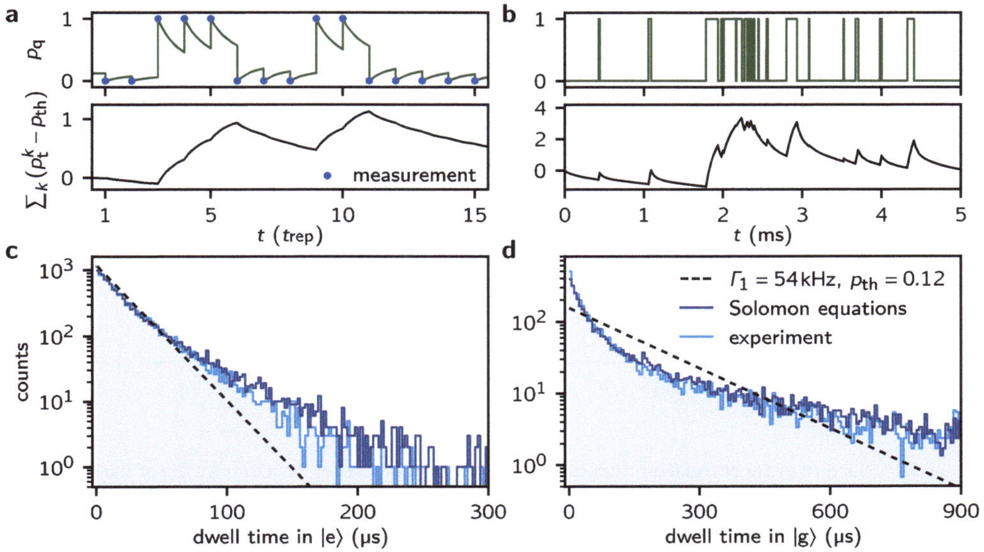

Figure 3.18: Comparison between Solomon-equations-simulated and measured quantum jumps. **a** Schematic simulation of a quantum jump trace using the Solomon equations. The blue points show the positions and outcomes of the measurement at t_{rep} intervals. The qubit is projected to its eigenstates with the probability corresponding to the calculated qubit population p_q. In the lower panel we show the excess population of the TLSs. It is assumed that the population of the TLSs remains unchanged by the measurement process. In between the measurements the system evolves according to the Solomon equations (Eq. 3.77). **b** Simulation of a longer quantum jump trace following the same method as in panel a with the device parameters from Ref. [21]. Here, only the qubit and excess TLS populations are shown at the time of the measurement. Due to the heat capacity of the environment and the relatively low thermal population $p_{th} = 0.12$, the quantum jumps appear bunched when the TLS temperature is increased. **c, d** Measured and Solomon-equations-simulated quantum jump statistics of the qubit dwell times in $|e\rangle$ and $|g\rangle$, respectively. The black dashed line shows the expected exponential distribution for a Born-Markovian qubit environment with qubit relaxation $\Gamma_1 = \Gamma_q + \Gamma_{TLSs}$ in thermal equilibrium corresponding to p_{th}. The histograms are generated as described in Fig. 3.12. Reproduced from Ref. [86].

The quantum jump statistics generated in this way is contrasted in Fig. 3.18c, d with the experiment and shows quantitatively an even better agreement than the SSE in Fig. 3.17.

Importantly, it should be noted that measuring non-Poissonian quantum jump statistics hampers the accurate extraction of the qubit's transition rates $\Gamma_{\uparrow,\downarrow}$. In thermal equilibrium, the extraction of the transition rates as discussed in Sec. 2.7.4 is equivalent to determining the inverse of the average qubit dwell times. As a consequence of the non-exponential distributions the qubit lifetime is overestimated. In Sec. 2.7.5, a qubit lifetime of T_1 = 21.5 μs was extracted, while here T_1 = 18.5 μs was used for the quantum jump simulations. The effect is even more drastic in

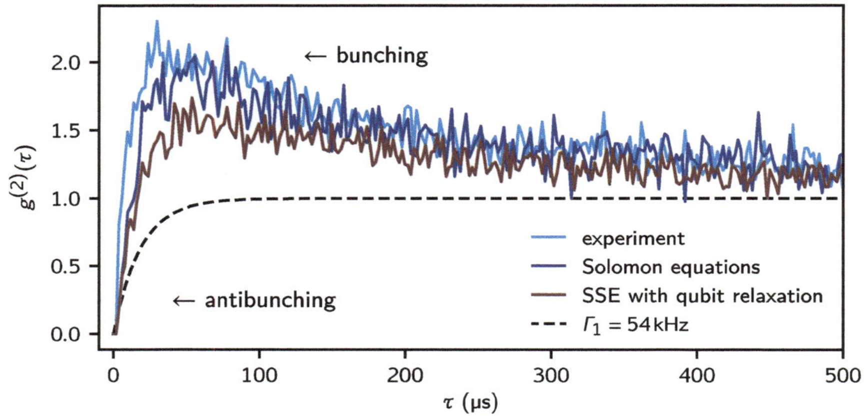

Figure 3.19: Second order correlation function. Second order correlation function $g^{(2)}(\tau)$ of the measured and simulated quantum jump traces shown in Fig. 3.17 and Fig. 3.18. The dashed black curve shows the correlation function for a Born-Markovian environment.

Fig. 3.21. Here, the rates $T_\uparrow = 71.6\,\mu s$ and $T_\downarrow = 6.2\,\mu s$ have been measured and the rates $T_\uparrow = 52.1\,\mu s$ and $T_\downarrow = 4.3\,\mu s$ have been used in the simulation to reproduce the measured statistics. The qubit lifetime can only be extracted precisely when the TLSs are in a well defined state, meaning that the TLSs fluctuate independently around their mean population. A well defined reservoir state is given after the TLS hyperpolarization. The increased relaxation rate Γ_1 that can be seen at the beginning of the measurements shown in Fig. 2.20b can be modeled when the rates are extracted from quantum jump traces simulated by the Solomon equations.

So far we only looked at the statistics of the qubit dwell times, which are also known as the waiting time distributions [136]. There are of course numerous other statistics that could be investigated and could potentially better tell apart the quantum jump traces generated by the SSE or Solomon equations. In Fig. 3.17a and Fig. 3.18b, it can be seen that the rate of quantum jumps increases with increasing energy and, more precisely, with increasing temperature in the system. The highest rate will be observed when the qubit experiences infinite temperature. When we interpret the quantum jump from the excited state to the ground state as the result of a photon emission, the second order correlation function $g^{(2)}(\tau) = \langle I(t+\tau)I(t)\rangle / \langle I(t)\rangle\langle I(t+\tau)\rangle$ can be computed from the intensity $I(t) \in \{0,1\}$ of emitted photons [166–168]. The correlation functions are depicted in Fig. 3.19. Since the qubit is a single photon emitter the photons must be antibunching on short time scales. However, for longer times, one finds $g^{(2)}(\tau) > 1$, which indicates thermal light, i.e. photon bunching. The correlation function for a Born-Markovian qubit environment is shown as well. Here, the correlation function is simply $g^{(2)}(\tau) = 1 - e^{-\Gamma_1\tau} < 1$, since the emittance of a photon is proportional to the qubit population.

3.6.6 Fluctuations in the TLS environment

Besides the occasional reconfiguration of the TLS environment presented in Sec. 2.7.1, the system can also undergo rapid fluctuations, which can last for longer periods of time. These fluctuations can be well studied by measuring quantum jump distributions, which can be done in a few seconds [53].

Fig. 3.20a shows the qubit lifetime that has been extracted from the quantum jumps distributions of the various interleaved relaxation measurements shown in Fig. 2.16 and Fig. 2.17. Except where the qubit lifetime jumps to $\sim 30\,\mu s$, the quantum jump distributions all show the non-exponential behavior that can be seen in Fig. 3.18 and Fig. 3.17, showing the quantum jump distribution at the position of the black arrow. The lifetime is determined from the average qubit dwell times $T_\uparrow = \langle t_{|g\rangle} \rangle = 1/\Gamma_\uparrow$ and $T_\downarrow = \langle t_{|e\rangle} \rangle = 1/\Gamma_\downarrow$ and using $T_1 = 1/(\Gamma_\uparrow + \Gamma_\downarrow)$. It should be noted that the extraction of the transition rates as discussed in Sec. 2.7.4 is an average over all iterations. The fluctuations of the qubit lifetime can therefore easily be overseen. If the qubit environment is stable and in thermal equilibrium, both approaches yield the same qubit lifetime. This comparably stable qubit environment is certainly one of the reasons for the successful modeling of the relaxation measurements in Sec. 2.7.2.

Fig. 3.20bc shows the situation of rapid fluctuations of the quantum jump distributions and qubit lifetime. Interestingly, when the qubit lifetime is short, in particular the quantum jump statistics of the ground state becomes highly non-exponential, and when the qubit lifetime is long, the statistics becomes exponential, except for the very first few bins in the ground state distribution. For lifetimes comparable to the ones extracted in Fig. 3.20a, the distributions (blue histogram in Fig. 3.20c) are similar to those measured in Fig. 3.20a.

These fluctuations between Poissonian and non-Poissonian statistics have been observed before on a junction array fluxonium with very similar time-scales [53]. It has been speculated that these fluctuations originate from a fluctuating quasiparticle background. Periods where the number of excess quasiparticles is stable would show a Poissonian statistics and have been denoted as quiet, whereas periods with a fluctuating number of excess quasiparticles would yield a non-Poissonian statistics and have been denoted as noisy. However, there are a few indications in Ref. [53] that rather suggest the presence of a two-level system environment in addition to excess quasiparticles that certainly exist, in particular after a strong quasiparticle generation pulse. First of all, it can be seen in the Supplementary of Ref. [53] that the qubit temperature after the generation pulse relaxes slowly with millisecond time-scales, similar to the findings in Fig. 2.16. Strikingly, the qubit temperature does not relax exponentially to the idle temperature in the absence of quasiparticle generation pulses but relaxes to higher temperatures. In principle, this could be explained by the microwave periphery that is heated up by all these strong microwave quasiparticle generation pulses. Furthermore, a rather low quasiparticle density was extracted during the noisy period, which remained an open question. Even more revealing, measurements on the same device reported in Ref. [71] show, after qubit excitation

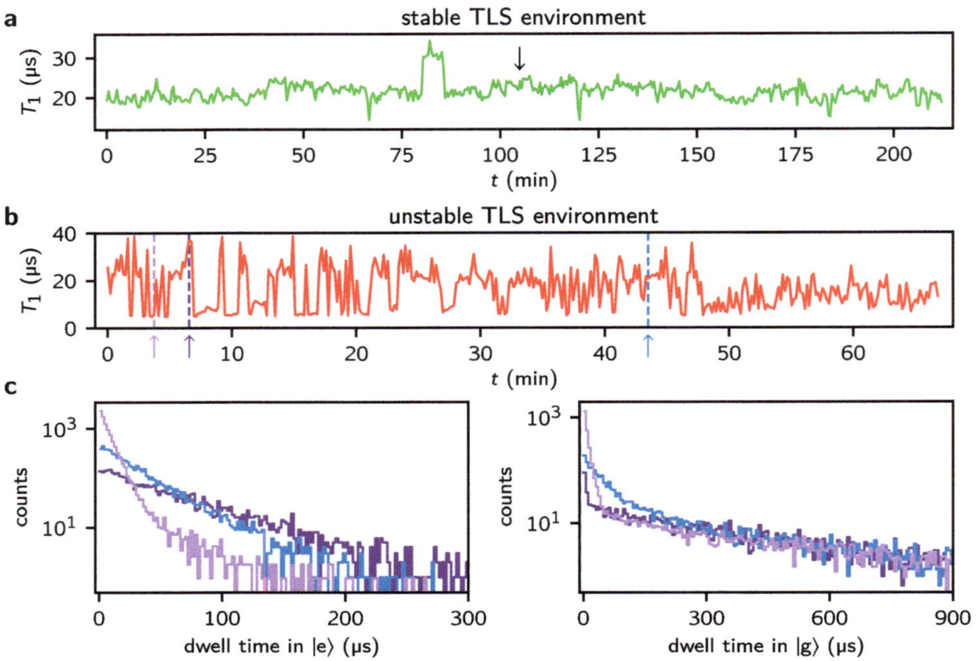

Figure 3.20: Fluctuations in the TLS environment. a Comparably stable qubit lifetime extracted from the various interleaved relaxation measurements shown in Fig. 2.16 and Fig. 2.17. Here, the lifetime is determined from the average qubit dwell times of the quantum jump statistics. In thermal equilibrium this approach coincides with the method presented in 2.7.4. The black arrow shows the position of the quantum jump statistics depicted in Fig. 3.18 and Fig. 3.17. The time resolution is $\sim$ 25 s, partially limited by the data transfer (Sec. 2.7.2). **b** Comparably unstable qubit lifetime, measured every $\sim$ 10 s by recording a 1.5 s long stroboscopic quantum jump trace with a repetition time of t_{rep} = 1.5 μs. The data was recorded in same cooldown eight days prior to the measurements shown in panel a. **c** Quantum jump statistics for different qubit lifetimes indicated by the arrows in panel b. Interestingly, when T_1 is short, in particular the quantum jump statistics of the ground state becomes non-exponential and when T_1 is large the statistics becomes exponential. The blue histogram was chosen to be similar to the one shown in Fig. 2.16 and Fig. 2.17. The unstable qubit behavior resembles the fluctuations reported in Ref. [53] measured on a junction array fluxonium. In this publication, Poissonian and non-Poissonian statistics have been denoted as quiet and noisy behavior, respectively.

with a π-pulse, fluctuating bi-exponential relaxations with a fast component on the order of hundreds of microseconds and a slow component on the order of 1-2 ms. In measurements following a saturation pulse, single exponential relaxations of up to 8 ms have been measured. This was interpreted as a fluctuating quasiparticle background, however, a long-lived TLS environment as reported here could be an alternative explanation. In the following, it will be shown that the fluctuations in Fig. 3.20bc arise from the most resonant TLS.

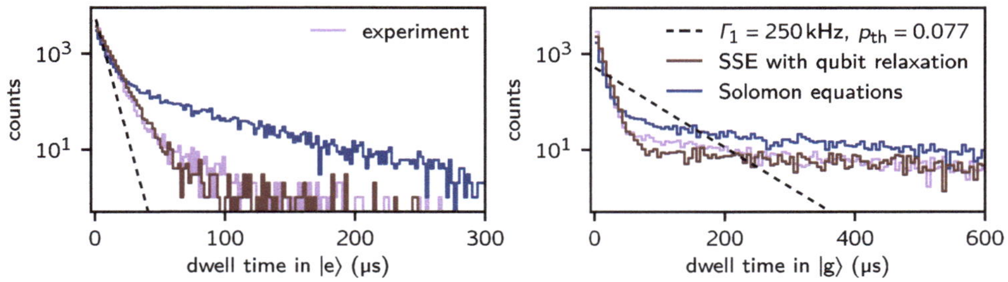

Figure 3.21: Observation of measurement back action. Quantum jump statistics of the qubit with the strongly coupled resonant TLS taken from Fig. 3.20c. The Solomon equations can not reproduce the measured quantum jump statistics, in contrast to the SSE, which considers the measurement back action. For further details see text.

3.6.7 Observation of measurement back action

The fluctuating qubit lifetime to larger values reaching ~ 35 µs with a corresponding exponential quantum jump distribution (Fig. 3.20bc) can only be explained by a decoupling of the TLS environment. The reason is that a fluctuating intrinsic qubit environment can not improve the qubit lifetime to ~ 35 µs, and moreover, the TLSs would still give a non-exponential quantum jump statistics. Interestingly, a complete decoupling of the most resonant TLS yields precisely the 35 µs lifetime. In addition, the decoupling of the most resonant TLS can also explain the exponential quantum jump distribution, since the non-exponential statistics comes mainly from the dominant TLS (Sec. 3.6.5). The logical consequence is that the extracted qubit lifetime of 5.7 µs is the result of a strongly increased coupling between the most resonant TLS and the qubit. Assuming a mutual decoherence of Γ_2 = 1.5 MHz, a coupling of $2\pi \cdot 53$ kHz is needed in contrast to the $2\pi \cdot 22$ kHz-coupling of all the other TLSs (Sec. 2.8). With this strongly coupled TLS, the Solomon equations fail to qualitatively reproduce the measured quantum jump statistics, as can be seen in Fig. 3.21. The simulated statistics is unaffected by the number of TLSs that are included in addition to the most resonant TLS. In contrast, the lossy SSE still reproduces the measured highly non-exponential quantum jump statistics because it accounts for the measurement back action. The SSE simulation was simply performed for the qubit and the most resonant TLS. The qubit lifetime was adjusted to T_1 = 4.0 µs, which yields the coupling strength $2\pi \cdot 65$ kHz for the most resonant TLS. When a lower TLS decoherence of Γ_2^t = 0.1 MHz is assumed, the extracted qubit lifetime increases due to the quantum Zeno effect, since the measurement repetition time t_{rep} = 1.5 µs is shorter than the mutual decoherence. To compensate the quantum Zeno effect, the qubit life time has to be further decreased in the simulation.

In conclusion, what we see here is indirect evidence of the measurement back action on the TLS. The fact that the qubit and the TLS get highly entangled during their free evolution after the projective measurement was detected even though the qubit and the TLS are relatively weakly coupled and exposed to strong dephasing. It is the energy that is shuffled around by the measurement process that makes the difference in the measured statistics. Since with increasing numbers of TLSs the statistics can be described with the Solomon equations, which neglect the measurement back action, we observed here a quantum-to-classical transition, in the sense that superpositions and measurement back action are not observed in the classical world.

4 Conclusion

The aim of this work was to implement granular aluminum fluxoniums and to investigate their performance. The presented fluxonium was one of the first implementations using disordered high kinetic inductance materials. Meanwhile, several other high kinetic inductance materials have been investigated and are used in various kinds of quantum devices [169, 170]. Still, granular aluminum stands out for its very high tunable kinetic inductance, its magnetic field resilience [48, 171] and its compatibility with the standard shadow-evaporation fabrication technique. The first noteworthy peculiarity that we routinely observe in our granular aluminum fluxoniums is the resilience to large photon numbers during the qubit readout. The reason for this photon resilience is likely a consequence of the cleaner mode structure of the granular aluminum superinductor in contrast to conventional high impedance quantum circuits that are implemented with junction arrays. With an appropriate meandering of the superinductor, very compact designs can be implemented, which further lifts the spurious modes far beyond the operational frequency range and allows the implementation of ultra high impedance circuits. When it comes to arguably the most important figure of merit for superconducting qubits, the qubit lifetime, we have measured values of up to a hundred microseconds, which is an order of magnitude lower than state-of-the-art fluxonium qubits implemented with junction arrays. Thus, the objective was to find the dominant loss mechanism. Extrapolating from resonator measurements and the overall flux dependence of the qubit lifetime, inductive loss from trapped quasiparticles in granular aluminum was expected to be the limiting loss mechanism of our granular aluminum devices. This notion was in line with further experiments that could reproduce previous findings in the field, showing that a sequence of saturation pulses could alter the qubit relaxation dynamics, which was interpreted as an enhanced quasiparticle diffusion [78].

The first part of this work dealt with the implementation of a quantum Szilard engine using active feedback control to drastically increase the interaction of the qubit with its environment. To date, the quantum Szilard engine presented in this work is one of the most complete experimental realizations of its kind described in the literature. Using the quantum Szilard engine, we were able to hyperpolarize the qubit environment and could show that the qubit is coupled to a heretofore unknown two-level system (TLS) environment. Interestingly, the qubit and the TLSs are each other's dominant loss mechanism. Limited only by the qubit preparation fidelity, the achievable TLS hyperpolarization is close to unity as a consequence of the, in our experiment immeasurable, long intrinsic lifetime of the TLSs exceeding 50 ms.

Concomitantly, the qubit's population exhibits remarkably long and non-exponential relaxation dynamics. The presented work therefore sheds new light on previous experiments and showcases the challenges and pitfalls of extracting T_1 from qubit population relaxation data. By extracting T_1 from stroboscopic quantum jump traces, we show experimentally that the qubit lifetime is unchanged for negative, infinite, and cryogenic temperatures, thereby ruling out the enhanced quasiparticle diffusion hypothesis, at least for our qubit. The often proposed idea of increasing the qubit lifetime by saturating the TLS must also be clearly rejected. The only certain way to increase the qubit lifetime is to remove the TLS environment.

The second part of this work dealt mainly with the theoretical description of the qubit relaxation in the presence of a long-lived TLS environment. An infinite number of weakly coupled TLSs provides a Born-Markovian environment for the qubit. In this case, the well known Bloch-Redfield master equation is the method of choice to describe relaxation and decoherence of the qubit. However, the situation changes drastically for a finite number of long-lived TLSs, as the Born approximation is no longer valid. This situation is typically found in NMR experiments. The underlying dynamics of the qubit and the TLSs may be interpreted as the nuclear Overhauser effect. For this reason, a rigorous derivation of the Solomon equations has been presented, adapted to the field of quantum information and extended for an arbitrary number of TLSs. The accuracy of the Solomon equations is demonstrated by describing the measured qubit relaxation on time scales more than three orders of magnitude longer than the qubit lifetime. On long time scales, the relaxation is described by a power-law, which was deduced from the Solomon equations and confirmed experimentally. In future work, it would be interesting to derive expressions for the TLS induced qubit decoherence and Lamb shift, analogous to the derivation of the Solomon equations.

Since the average relaxation, described by the Solomon equations, hides the quantum mechanical complexity, we asked ourselves whether the underlying stochastic dynamics that is essential for the cross-relaxation could be uncovered and understood by investigating the real-time dynamics of the stroboscopic quantum jump traces. To our surprise, we found both long periods of stable non-exponential quantum jump distributions and periods where the quantum jump statistics undergoes rapid fluctuations between highly non-Poissonian statistics with short T_1 and Poissonian statistics with longer qubit lifetime. A very similar fluctuating behavior has been observed before on junction array fluxoniums and was attributed to a fluctuating quasiparticle background [53]. Using the diffusive stochastic Schrödinger equation, we were able to reproduce the highly non-Poissonian quantum jump distribution and to show that the fluctuations are caused by the most resonant TLS. Thus, the coupling strength between the qubit and the TLSs can take on values as high as $\sim 65\,\mathrm{kHz}$, which could potentially also be larger when a higher TLS dephasing is assumed. Moreover, for this rather strong interaction, entanglement and measurement back action are essential to reproduced the measured distribution.

Interestingly, with increasing system size, and as long as individual TLSs do not completely dominate the dynamics, we show that the Solomon equations can also be used to describe the measured non-Poissonian quantum jump statistics. Since this procedure neglects entanglement and measurement back action but can still reproduce the measured statistics, it indicates that there is a transition after which local measurements on the qubit can no longer distinguish between quantum mechanical and classical statistics, which can be viewed as a quantum-to-classical transition. The conditions that allow for the modeling of the quantum jumps with the Solomon equations need to be further investigated. Finally, different stochastic unravelings of the system dynamics should be studied to see which one of these can reproduce the measured statistics.

Outlook

Certainly, the next important step is to identify the physical origin of these long-lived TLSs. One the one hand, the long lifetime of the TLSs suggests that the TLSs are spins. Since the TLS environment could be observed at various frequencies, the TLSs must be composed of at least two spins that are coupled together, e.g. via the dipolar interaction. A spatial distribution of the spins would then translate into a frequency distribution. Indeed, such a broad spin frequency distribution can be observed in experiments [172]. On the other hand, the coupling strength that we extracted in our experiments is rather large compared to the coupling that was achieved in Ref. [3] with an ultra low impedance resonator. This suggests that the TLSs are trapped quasiparticles, Shiba spins or other subgap states in the superconductor. It could certainly be that the disordered nature of granular aluminum is susceptible to such environments, however it would be very unlikely if this environment is completely absent in junction array fluxoniums. The observation of similar effects in the past and further experiments that have been presented to us seem to indicate that these long-lived TLSs are ubiquitous to superconducting hardware.

Therefore, the following experiments are suggested. Using a fast-flux tunable fluxonium, the Szilard engine may be operated at various frequencies. In this case, the active feedback is no longer needed and can be replaced by a fast flux reset. This experiment could reveal the frequency distribution of the TLSs. In addition, the flux tunability would allow mapping out the spectral hole that was burned into the TLS environment. In this way, the decoherence and the lifetime of the TLSs can be determined. The first steps in this direction have already been undertaken. The spectral hole burning and lifetime measurements have recently been presented on a dielectric TLS environment using the methods developed in this thesis [173].

In order to investigate a possible magnetic field susceptibility of the TLSs, the fluxonium can be placed in an in-plane magnetic field. Fortunately, the magnetic field can be rather moderate, as the spin transitions only need to be lifted above the qubit frequency. One may also place a few spins on purpose in the vicinity of the superinductor to investigate in more detail the underlying stochastic dynamics as well as cooperative effects.

Part II

Detailed information

5 Quantum circuits

In this chapter, the quantization of superconducting circuits is introduced. The circuit quantization is detailed for various circuits that have been encountered over the course of this thesis. The main focus is to provide insights into the characteristics of junction arrays and granular aluminum (grAl) and the quantization of various flux type qubits. The calculations are compared to experiments performed with grAl fluxoniums.

5.1 Circuit quantization

The quantization of superconducting circuits has been discussed in numerous publications [174–177] and has gained renewed interested in the last years [178–183]. Certainly the most widely used circuit quantization technique is the canonical quantization that was introduced by Devoret [175], which was further worked out by Vool [176]. One of the main motivations was to develop a procedure that makes no difference between superconducting elements and normal conducting cavities or waveguides. For this reason, their recipe is straightforward to apply and has become the core of automatized circuit quantization software packages [184, 185]. However, by treating the gauge invariant flux and the flux of the superconducting wave function on equal footing, the method has the drawback of not yielding the correct absolute ground state energy of the system. For the very same reason, analytical expressions of the Hamiltonian are not necessarily unique and may result in more complicated expressions. Even more importantly, for time-dependent external flux biases, additional terms must be included to yield correct predictions, as was demonstrated in a recent experiment [186].

The canonical quantization, i.e. the derivation of the Hamiltonian from a classical Lagrangian, is certainly the first obstacle one encounters, since one is used to directly setting up the Hamiltonian for all the elements in the system. In addition, one may be concerned by Groenewold's theorem [187], which states that the canonical quantization can be ambiguous. Similarly, we are forced to treat the kinetic non-linear Josephson energy in the Lagrangian as the potential energy. If we had to include also nonlinear capacitance terms, predicted to play a role for very small capacitances [188], the canonical quantization would simply not be viable, as the Legendre transformation could not be performed uniquely. Fortunately, there is always the option to fall back on a microscopic theory, as demonstrated by the circuit quantization of Eckern long before the first experimental observation of coherence in superconducting qubits [188].

There is yet another complication in the quantization of circuits. A Hamiltonian must consist solely of independent degrees of freedom, i.e. generalized coordinates. For instance, connecting inductors in a node while neglecting its capacitance to other nodes removes a degree of freedom but leaves behind a constraint. These constraints must be removed beforehand, either by deriving a Lagrangian with generalized coordinates or by means of a symplectic transformation [181]. Nonlinear constraints are discussed in [182] and can be removed by the Born-Oppenheimer approximation [189]. An illustrative example is given in Sec. 5.2.1.

In the following, the circuit quantization is motivated from a microscopic perspective by considering the superconducting condensate. The energy contributions of all basic elements are derived before proceeding with increasingly complex circuits.

The superconducting condensate

A microscopic derivation of superconducting circuits would ideally begin similar to to the Jellium model. The system Hamiltonian would include all the electrons in the circuit, or at least the condensed Cooper pairs, the positive background field of the nuclei, which confines the electrons to the circuit, and moreover it should also include the quantized electromagnetic field. With this Hamiltonian, it should even be possible to study quantum fields, i.e. the quantization of distributed circuits such as the long junction qubit [190, 191]. Since such a derivation is by far out of the scope of this thesis, a conceptually similar approach of greatly reduced complexity is presented instead.

Close to zero temperature one may use the Gross-Pitaevskii equation [192, 193] to describe the condensate of N repulsive Cooper pairs. The model Hamiltonian can be written as

$$H = \sum_{i=1}^{N} \frac{(\mathbf{p}_i - q\mathbf{A})^2}{2m} + U(\mathbf{r}_i) + \sum_{i<j} g\delta(\mathbf{r}_i - \mathbf{r}_j). \tag{5.1}$$

Here, $\mathbf{p}$ is the momentum operator of the Cooper pairs with mass $m = 2m_\mathrm{e}$ and charge $q = -2e$. The magnetic vector potential is encoded in $\mathbf{A}$. The potential U may first of all be used to confine the condensate in the superconducting circuit, but also to model Josephson junction tunnel barriers as well as the complex structure of granular aluminum. Finally, the repulsive interaction with coupling strength g is modeled as a contact potential.

Since the weakly interacting Cooper pairs will largely condensate in the same state, except for a small fraction called the quantum depletion [194], we can approximate the wave function as

$$\Psi = \psi(\mathbf{r}_1)\psi(\mathbf{r}_2)\ldots\psi(\mathbf{r}_N). \tag{5.2}$$

Within this Hartree-Fock approximation and following the Rayleigh-Ritz method (see also Sec. 5.3.4), the eigenfunctions and corresponding eigenenergies of the Hamiltonian in Eq. 5.1 are best approximated by solutions of the Gross-Pitaevskii equation

$$h\psi(\mathbf{r}) = \left(\frac{(\mathbf{p} - q\mathbf{A})^2}{2m} + U(\mathbf{r}) + g(N-1)|\psi(\mathbf{r})|^2\right)\psi(\mathbf{r}) = \epsilon\psi(\mathbf{r}). \tag{5.3}$$

This equation is also derived within the Ginzburg-Landau theory together with additional equations for the superconducting current and the magnetic field [195]. The Gross-Pitaevskii equation is also known as the non-linear Schrödinger equation [196].

Given a single particle solution ψ with single particle energy ϵ, the energy of the entire condensate is $E = N\epsilon$. Consequently, adding or removing a Cooper pair from the condensate is linked to the energy ϵ, which therefore plays the role of a chemical potential. Since the condensate can be assumed to fill more or less homogeneously the superconducting circuit with volume V, we can transition to the rotating frame of the condensate via $\psi \to \psi e^{-i/\hbar\mu t}$ with $\mu = g(N-1)/V$, which essentially removes the interaction energy in Eq. 5.3.

For the excitation of the condensate, the eigenenergies $E = N \cdot \epsilon$ are relevant. As a short-cut one often defines $\psi = \sqrt{n_\mathrm{s}(\mathbf{r})}e^{i\varphi(\mathbf{r})}$ normalized to N Cooper pairs, which introduces the Cooper pair density $n_\mathrm{s}(\mathbf{r})$ and the superconducting phase $\varphi(\mathbf{r})$. Alternatively, one can instead multiply the Hamiltonian by N and use a normalized wave function.

Basic elements

We begin with the most basic element, an inductance formed by a superconducting wire with length l, width w and height h as shown in Fig. 5.1. We will for now assume that $U(\mathbf{r}) = 0$ inside the wire so that the condensate density is homogeneous. Thus, considering the axis along the wire and assuming for simplicity that the vector potential pointing along the wire is constant and amounts to A, the Cooper pairs of a moving condensate must be described by virtue of Eq. 5.3 by a plane wave $\psi(x) = e^{\frac{i}{\hbar}px}/\sqrt{V}$. The movement in the wire can only be enabled when it is shunted by a capacitor with capacitance C, which allows the accumulation of charges at the capacitor plates. Consequently, a condensate displacement by the distance x corresponds to the charge $q = -2e/l \cdot x$ per Cooper pair on the capacitor plates. This coordinate transformation directly defines the corresponding flux operator

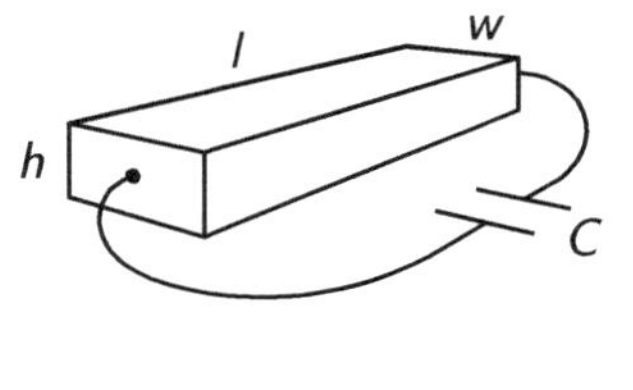

Figure 5.1: Superconducting wire with length l, width w and height h shunted by a capacitance C.

$\phi = -l/2e \cdot p$, so that $[q, \phi] = i\hbar$. The wave function can now be represented as $\psi(x) = e^{\frac{i}{\hbar}\phi q}/\sqrt{V}$ and even more instructive

$$\psi(x) = e^{-2\pi i \frac{\phi}{\Phi_0} \frac{x}{l}}/\sqrt{V}, \tag{5.4}$$

where the superconducting flux quantum $\Phi_0 = h/2e$ has been introduced. In order to treat ϕ as the position operator, one has to redefine the charge operator to be negative, i.e. $q := -q$, so that $[\phi, q] = i\hbar$.

In the Hamiltonian of the inductive wire shunted by the capacitance, the potential for the capacitor must be added, considering also offset charges q_g that can be induced by quasiparticles or gate voltages. The Hamiltonian $H = Nh$ then reads [176]

$$H = \frac{N}{2} \frac{(p + 2eA)^2}{2m_e} + \frac{(q - q_g)^2}{2C} \tag{5.5}$$

$$= \frac{1}{2} \left(\frac{4e^2 n_s}{2m_e} \right) \left(\frac{wh}{l} \right) (\phi - lA)^2 + \frac{(q - q_g)^2}{2C} \tag{5.6}$$

$$= \frac{(\phi - \Phi_{ext})^2}{2L_{kin}} + \frac{(q - q_g)^2}{2C}. \tag{5.7}$$

In the second line, the specific kinetic inductance $l_{kin} = 2m_e/4e^2 n_s$ that arises from the momentum of the Cooper pairs can be identified. In the last line, the kinetic inductance $L_{kin} = l_{kin} l/wh$ was introduced as well as the magnetic flux $\Phi_{ext} = lA$, essentially as the integrated vector potential along the wire.

Since the electromagnetic field energy was not included in the derivation, the geometric inductance L_{geo} must be added to the total inductance of the wire $L = L_{kin} + L_{geo}$. In fact, for materials with a low kinetic inductance, the geometric inductance L_{geo} gives usually the dominant contribution. For very sophisticated designs, the kinetic inductance can even be negligible [197]. In contrast, for high kinetic inductance materials such as granular aluminum, the geometric inductance can be neglected. Here, the increased kinetic inductance is not the result of a reduced electron density, but a consequence of the complex potential landscape $U(\mathbf{r}) > 0$. The condensate gains in potential energy when a phase gradient is imprinted on the condensate, similar to the effective electron mass in solid-state physics. In the dirty limit, valid for granular aluminum but also for thin film aluminum, it holds [17]

$$l_{kin} = \frac{\hbar\rho}{\pi\Delta}. \tag{5.8}$$

Here, ρ is the specific resistivity measured at the critical temperature T_c, and Δ is the superconducting gap at zero temperature. For granular aluminum, one can use in good approximation the room temperature resistivity, since the residual resistance ratio (RRR) is close to one [17, 198]. Alternatively, granular aluminum may also be modeled as a network of Josephson junctions, see Sec. 5.2.3.

After having introduced the role of inductors and capacitors, we now turn to the key element of superconducting circuits, the Josephson junction. A Josephson junction is a weak link between two superconductors, which enables the coherent tunneling of Cooper pairs across the junction, known as the Josephson effect. In conventional Josephson junctions of superconducting qubits, the tunnel barrier is implemented as an insulating oxide barrier. Since the barrier thickness is vanishingly small, on the order of $\sim 2\,\mathrm{nm}$ [199], the junction potential does not depend on the magnetic vector potential, and its kinetic energy contribution to the Hamiltonian is given by the potential [195]

$$V(\phi) = -E_\mathrm{J} \cos\left(\frac{2\pi}{\Phi_0}\phi\right), \tag{5.9}$$

with E_J being the Josephson energy and ϕ the flux difference of the superconducting condensate across the junction. In general, this potential energy contains higher order harmonics [200, 201], which have recently been observed in qubit spectroscopy measurements [199]. The derivative of $V(\phi)$ with respect to ϕ gives the current phase relation [201]

$$I(\phi) = I_\mathrm{c} \sin\left(\frac{2\pi}{\Phi_0}\phi\right), \tag{5.10}$$

with the critical current $I_\mathrm{c} = 2\pi/\Phi_0 \cdot E_\mathrm{J}$.

5.1.1 The superconducting ring

Certainly, the simplest superconducting circuit is the superconducting ring as shown in Fig. 5.2. It even comes with an anharmonic level structure. Unfortunately or fortunately, it is almost impossible to drive its transitions [202]. The superconducting ring is the essential part of gradiometric fluxoniums [203] (s. Sec. 5.3.1). The persistent current can be used as a local magnetic field bias [203, 204], but also as a magnetic field or non-linearity sensor in case of nonlinear inductors (s. Sec. 5.2.5). Following the previous section, the Hamiltonian of the superconducting ring with a potentially time-dependent external flux is

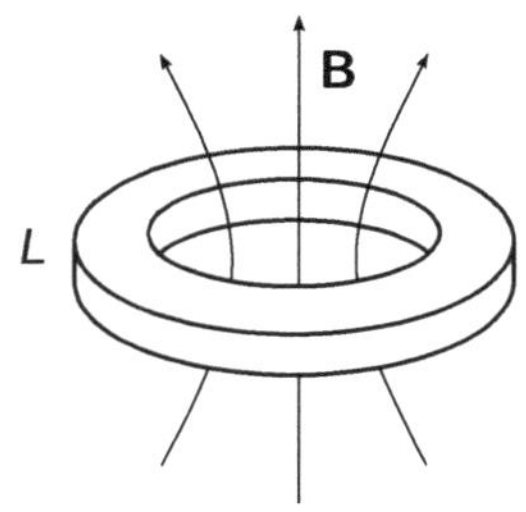

Figure 5.2: Superconducting ring with inductance L, here illustrated with a trapped flux and corresponding magnetic field **B**.

$$H = \frac{(\phi - \Phi_\mathrm{ext}(t))^2}{2L}. \tag{5.11}$$

Due to the periodic boundary condition, this Hamiltonian is solved by the wave functions

$$\psi_m(\vartheta) = \frac{1}{\sqrt{V}} e^{im\vartheta} e^{-\frac{i}{\hbar}\int \epsilon(t)\,\mathrm{d}t}. \tag{5.12}$$

Here, m is the number of trapped flux quanta and ϑ the angular coordinate. These wave functions belong to time-dependent energies

$$E_m = \frac{(m\Phi_0 - \Phi_{\mathrm{ext}}(t))^2}{2L} \tag{5.13}$$

that display the energy of the screening currents. As an example, suppose that the superonducting ring is made of granular aluminum with a kinetic inductance of $L = 500\,\mathrm{nH}$. In the absence of a magnetic field, the fundamental transition will therefore be

$$hf_{01} = \frac{\Phi_0^2}{2L} = 6.45\,\mathrm{GHz}. \tag{5.14}$$

Assuming a moderate resistivity of $l_{\mathrm{kin}} = 500\,\mathrm{pH/sq}$ and a wire width $w = 250\,\mathrm{nm}$, the ring consists of $s = 1000$ squares and has a diameter of 80 µm. For this geometry, we can calculate the additional geometric inductance. The ring may be imagined wrapped around a cylinder, for which an analytical formula is available[1]. For $s \gg 1$, it holds

$$L_{\mathrm{geo}} \approx \mu_0 r \left(\ln\left(\frac{8s}{2\pi}\right) - \frac{1}{2}\right) = 0.33\,\mathrm{nH}. \tag{5.15}$$

This shows that the geometric inductance in our fluxonium designs gives corrections in the percent level.

The usual argument for the endurance of persistent currents is that Cooper pairs are formed of electrons with opposite momentum. When Cooper pairs break apart and recombine they can not change the momentum of the condensate. Considering the electromagnetic environment, one might wonder why a persistent current can not be excited by incoming radiation as well as decay into its ground state by emitting a photon. For superconducting rings with a ring inductance in the nH-range, one can first of all argue that the transition frequency easily exceed the superconducting gap. Incoming radiation will thus rather break Cooper pairs apart, since in addition, the corresponding wavelengths are smaller than the dimensions of the ring. This argument is not sufficient for superconducting rings made out of granular aluminum. After all, it is the large angular momentum of the ring that can not be carried away

[1] The exact formula by Lorenz [205], here rearranged in Nagaoka's form [206], is $L_{\mathrm{geo}} = \frac{2\mu_0 r s}{3\pi} \frac{1}{\sqrt{1-k^2}} \left[\frac{1-k^2}{k^2}(K(k) - E(k)) + E(k) - k\right]$, where r is the radius of the cylinder, $k^2 = s^2/(s^2+\pi^2)$ with s being the number of squares around the cylinder jacket, and E, K denote the elliptic integrals.

by a single photon, in accordance with the physics of nearly stable nuclear spin isomers. Nonetheless, reliably driving transitions with large field amplitudes should in principle still be possible [207].

5.1.2 The charge qubit

The charge qubit is certainly the most widely used qubit type in superconducting quantum hardware. In particular, the transmon [208] with its reduced sensitivity to charge noise enables a multitude of sophisticated experiments [3, 7, 9, 62]. In this section, the theory of the charge qubit is briefly reviewed [208] to lay the foundations for the upcoming sections.

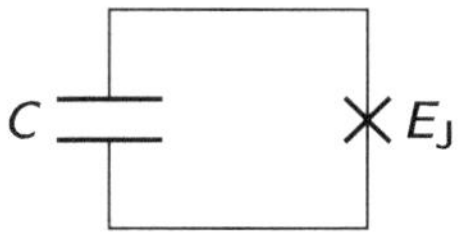

Figure 5.3: Equivalent circuit of the charge qubit with capacitance C and Josephson energy E_J.

The equivalent circuit of the charge qubit simply consists of a Josephson junction with Josephson energy E_J that is shunted by a capacitor with capacitance C, as depicted in Fig. 5.3. The capacitance is partly formed by the intrinsic capacitance of the junction as well as by an additional shunting capacitance, which allows to independently adjust the qubit frequency and sensitivity to charge noise, and gives the qubit its required dipole moment. The Hamiltonian reads

$$H = \frac{1}{2C}(q - q_\mathrm{g})^2 + E_\mathrm{J} \cos\left(\frac{2\pi}{\Phi_0}\phi\right). \tag{5.16}$$

In this form, the Hamiltonian is not very practical. A more convenient Hamiltonian can be obtained by transitioning into a scaled coordinate system, i.e. introducing new operators of the form $x = \phi/s$ and correspondingly $p = s \cdot q$. For instance, the potential energy may be visualized in units of flux quanta, hence $s = \Phi_0$. An even more compact Hamiltonian is obtained for $s = \Phi_0/2\pi$. This introduces the phase operator $\varphi := x = \phi/s$ and the number operator $n := p/\hbar = s \cdot q/\hbar$. Since the reduced Planck constant was absorbed in the number operator, we now have $n = -i\partial/\partial\varphi$, or equivalently $[\varphi, n] = i$. Using these dimensionless operators, the Hamiltonian takes on the form

$$H = \frac{1}{2}E_\mathrm{C}(n - n_\mathrm{g})^2 + E_\mathrm{J} \cos(\varphi). \tag{5.17}$$

Here, we introduced the charging energy $E_\mathrm{C} = 4e^2/C$. Note that the nowadays more common definition of $E_\mathrm{C} = e^2/2C$ is not used here, for good reasons[2].

[2]The definition $E_\mathrm{C} = e^2/2C$ stems historically from single electron transistors [209] and reminds us that Cooper pairs are formed by two electrons. For consistency, we should then also use the flux quantum and resistance quantum of the electron. The definition used here is consistent with the usual definition of the inductive energy E_L. For E_L, we have no other choice, since it has to coincide with E_J,

Yet another scaling should be mentioned that is practical when comparing different Hamiltonians. In fact, for the comparison of multidimensional Hamiltonians this scaling is even mandatory (s. Sec. 6.3). The idea is to essentially rescale the Hamiltonian in Eq. 5.17 even further. Using $s = \Phi_0/2\pi \cdot E_{\mathrm{C}}^{1/2}$ we obtain

$$H = \frac{1}{2}\frac{(p - p_{\mathrm{g}})^2}{\hbar^2} - E_{\mathrm{J}} \cos\left(\sqrt{E_{\mathrm{C}}} x\right) \tag{5.18}$$

$$= \frac{1}{2}\frac{(p - p_{\mathrm{g}})^2}{\hbar^2} - E_{\mathrm{J}} + \frac{\hbar^2\omega^2}{2}x^2 - \frac{E_{\mathrm{C}}}{24}x^4 + \ldots, \tag{5.19}$$

where we can identify the angular frequency ω of the harmonic oscillator via $\hbar\omega = \sqrt{E_{\mathrm{J}}E_{\mathrm{C}}}$. With this scaling, all Hamiltonians describe a particle with mass $m = \hbar^2$ and all information of the spectrum is encoded in the potential.

Charge dispersion

Since the phase difference of the condensates on each of the islands can only be between $-\pi$ and π, the potential energy as well as the wave functions should be imagined wrapped around a cylinder. Mathematically speaking, the wave functions are subject to the periodic boundary condition

$$\psi(\varphi) = \psi(\varphi + 2\pi). \tag{5.20}$$

This boundary condition is responsible for giving discrete energy levels, but also making the qubit charge sensitive.

For a constant or slowly drifting offset charge in time, we can make the ansatz $\psi(\varphi) = e^{i n_{\mathrm{g}} \varphi}\psi'(\varphi)$, or equivalently, we can introduce the Hamiltonian $H' = e^{-i n_{\mathrm{g}}\varphi} H e^{i n_{\mathrm{g}}\varphi}$ for the wave function ψ'. We obtain

$$H' = \frac{1}{2}E_{\mathrm{C}}n^2 + E_{\mathrm{J}}\cos(\varphi). \tag{5.21}$$

The corresponding eigenfunctions ψ'_{nk} with eigenenergies $E_{n,k}$ are known as Mathieu functions. Due to the periodic potential landscape they are of the form

$$\psi'_{nk}(\varphi) = e^{ik\varphi}u_{nk}(\varphi) \tag{5.22}$$

with the wave vector k restricted to the first Brillouin zone and $u_{nk}(\varphi)$ being periodic with the same periodicity as the potential. The boundary condition Eq. 5.20 is satisfied when we choose $k = -n_{\mathrm{g}}$, yielding the eigenfunctions

$$\psi_{n,n_{\mathrm{g}}}(\varphi) = e^{i(n_{\mathrm{g}}+k)\varphi}u_{nk}(\varphi) = u_{n,-n_{\mathrm{g}}}(\varphi) \tag{5.23}$$

where the prefactor of one-half arises from the Taylor expansion of the cosine. Using E_{C} as defined here yields more symmetric and compact formulas. Why should we drag along a factor of eight in all relevant calculations and formulas?

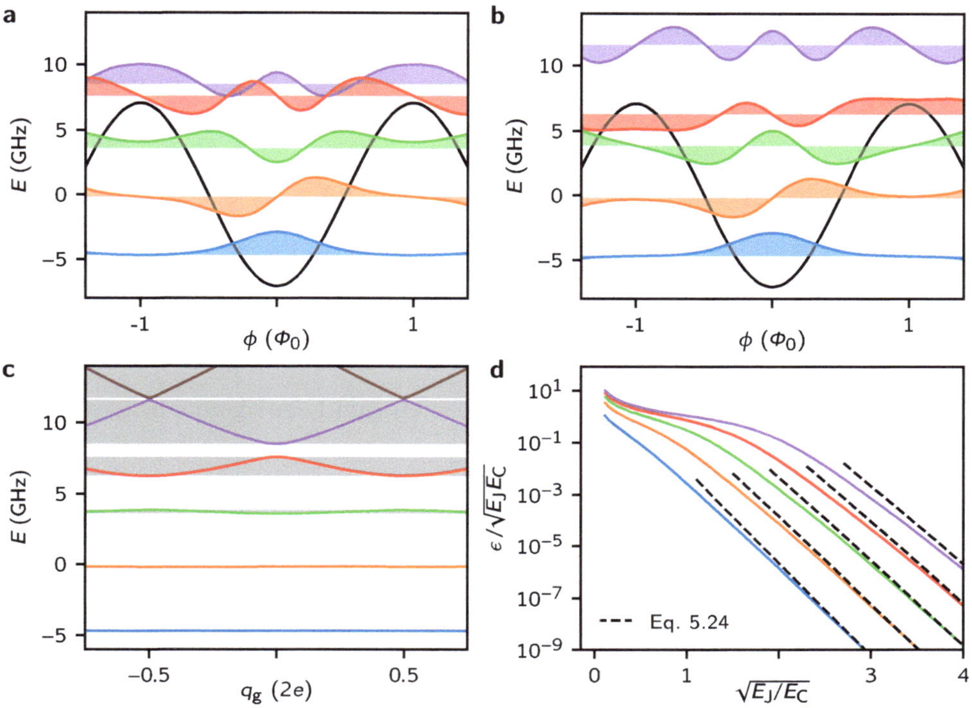

Figure 5.4: Spectrum of the charge qubit. Wave functions ψ'_{n,n_g} of the charge qubit with $\sqrt{E_\mathrm{J}E_\mathrm{C}} = 5\,\mathrm{GHz}$ and $\sqrt{E_\mathrm{J}/E_\mathrm{C}} = \sqrt{2}$ for an **a** even and **b** odd number of offset electrons. **c** Spectrum as a function of the offset charge. The Bloch bands are indicated in light gray. **d** Charge dispersion ϵ_n as a function of $\sqrt{E_\mathrm{J}/E_\mathrm{C}}$. The approximation formula is depicted by the dashed black line.

with corresponding energies E_{n,n_g} that are offset charge dependent.

The energies $E_{n,k}$ are dense in k and form separated Bloch bands. For $E_\mathrm{J}/E_\mathrm{C} > 1$, the energy width ϵ of the lower Bloch bands can be approximated by [208]

$$\epsilon_n = \frac{\sqrt{E_\mathrm{J}E_\mathrm{C}}}{\sqrt{\pi}} \frac{2^{5n+4}}{n!} \sqrt[4]{\frac{E_\mathrm{J}}{E_\mathrm{C}}}^{2n+1} e^{-8\sqrt{\frac{E_\mathrm{J}}{E_\mathrm{C}}}}. \tag{5.24}$$

In practice, deviations of the sinusoidal current phase relation can lead to amplification of the charge dispersion [199]. Finally, the energy spread of the lowest band can be used as a rough approximation for the fluxonium qubit frequency (s. Sec. 5.3.5).

The wave functions and their charge dispersion are depicted in Fig. 5.4 for the lower levels. The numerical approximation of the Mathieu functions is discussed in the next paragraph.

Numerical diagonalization

A convenient and fast method for diagonalizing the Hamiltonian in Eq. 5.17, which exploits the periodic symmetry, is to develop the wave function in the charge basis:

$$\chi_m(\varphi) = \langle \varphi | \chi_m \rangle = \frac{e^{im\varphi}}{\sqrt{2\pi}}. \tag{5.25}$$

In this basis, the matrix elements of the Hamiltonian are simply

$$\mathbf{H}_{m,n} = \langle \chi_m | H | \chi_n \rangle = \frac{E_\mathrm{C}}{2}(n - n_\mathrm{g})^2 \delta_{m,n} - \frac{E_\mathrm{J}}{2}(\delta_{m,n+1} + \delta_{m,n-1}), \tag{5.26}$$

which is a tridiagonal matrix structure that can efficiently be diagonalized numerically.

Since for $\sqrt{E_\mathrm{J}/E_\mathrm{C}} > 1$ the wave functions are well defined by the harmonic potential, one can use the wave functions of the harmonic oscillator to approximate the lower charge qubit states. The harmonic oscillator wave functions are of the form

$$h_n(x) = \langle x | h_n \rangle = \left(\frac{\beta^2}{\pi}\right)^{1/4} \frac{1}{\sqrt{2^n n!}} \mathrm{H}_n(\beta x) e^{-\frac{\beta^2 x^2}{2}}, \tag{5.27}$$

with H_n denoting the Hermite polynomials of order n. The parameter β depends on the scaling s. For the Hamiltonian in Eq. 5.17, we have $\beta = \sqrt[4]{E_\mathrm{J}/E_\mathrm{C}}$, while for the Hamiltonian in Eq. 5.18, we find $\beta = \sqrt{\hbar\omega}$.

The flux and charge operators can now be expressed with the creation and annihilation operators of the harmonic oscillator. It holds

$$x = \frac{1}{\sqrt{2}\beta}(a^\dagger + a), \qquad p = \frac{i\hbar\beta}{\sqrt{2}}(a^\dagger - a). \tag{5.28}$$

With these operators, the Hamiltonian in Eq. 5.19 becomes

$$H' \approx \hbar\omega \left(a^\dagger a + \frac{1}{2}\right) - E_\mathrm{J} - \frac{E_\mathrm{C}}{96}(a^\dagger + a)^4, \tag{5.29}$$

and in first order approximation we obtain the energies [208]

$$E_n \approx -E_\mathrm{J} + \hbar\omega \left(n + \frac{1}{2}\right) - \frac{E_\mathrm{C}}{16}\left(n^2 + n + \frac{1}{2}\right). \tag{5.30}$$

Note that the offset charge dependence is lost since harmonic oscillator states do not obey the periodic boundary condition.

5.2 On junction arrays and granular aluminum

In the following section, the transition from junction arrays to granular aluminum is discussed. The complexity showcases the need of the Born-Oppenheimer approximation that is introduced in the beginning with an illustrative example.

5.2.1 Two junctions in series

Nonlinear kinetic elements are often defined by their current phase relation. The current phase relation can then be integrated to yield the potential energy for the Hamiltonian up to a constant offset energy. A convenient method of modeling networks of non-linear kinetic elements is therefore to derive the current phase relation. However, this strategy does not always work [182]. For example, a junction in series with an linear inductor can have multiple solutions. In practice, the ambiguity of the solution branches is lifted when the charging energy of the individual elements is taken into account. However, this comes at the cost of an increased Hilbert space, which may be reduced using the Born-Oppenheimer approximation.

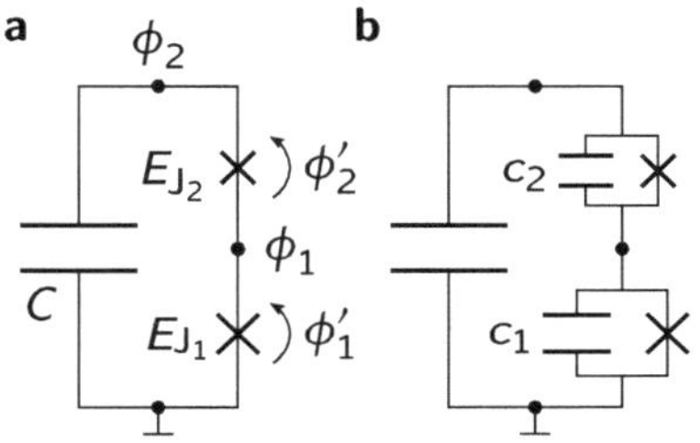

Figure 5.5: Two junctions in series and shunted by a capacitor **a** without and **b** with junction capacitors.

As an illustrative example, we show here the solution of two possibly distinct Josephson junctions in series that are shunted by a capacitor, as depicted in Fig. 5.5. In Ref. [210], we presented the analytical solution of the Hamiltonian via the current phase relation of both junctions. This derivation is briefly reviewed and compared to the Born-Oppenheimer approximation in the next paragraph.

The current phase relation of both junctions is obtained by solving

$$I_{c_1}\sin(\varphi_1) = I_{c_2}\sin(\varphi_2 - \varphi_1), \tag{5.31}$$

with I_{c_i} being the critical currents of the two junctions. The flux variables without prime denote the flux of the condensate with respect to the ground node flux, whereas the variables with prime denote the flux difference across the junctions (Fig. 5.5). The above equation has always two distinct solutions φ_1 and $\varphi_1 + \pi$. We have to choose the branch that is energetically favored, i.e. the branch where the currents are driven by the voltage. In order to choose the right branch, we can assume without loss of generality that $I_{c_1} > I_{c_2}$. Then, using the identity $\sin(x-y) = \sin(x)\cos(y) - \cos(x)\sin(y)$, we find

$$\varphi_1 = \arctan\left(\frac{\sin(\varphi_2)}{r + \cos(\varphi_2)}\right), \tag{5.32}$$

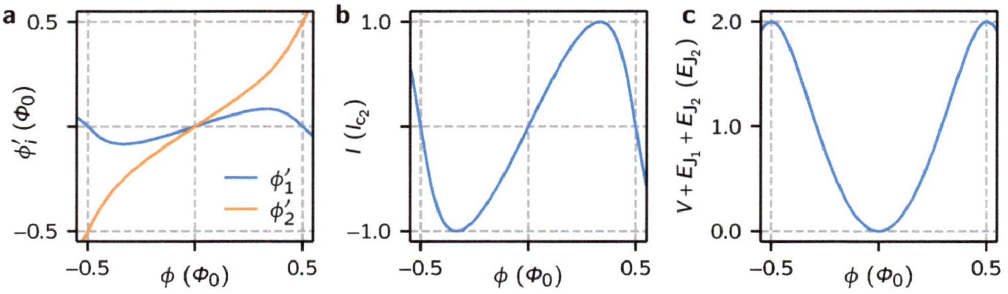

Figure 5.6: Two junctions in series. **a** Distribution of the total flux ϕ on both junctions. **b** Current phase relation of two junctions in series and **c** corresponding potential energy.

where we defined $r = I_{c_1}/I_{c_2} \geq 1$. The flux that drops at the junctions ϕ_i' is depicted in Fig. 5.6. Insertion of φ_1 in the current phase relation of the first junction yields the new current phase relation $I(\varphi)$ of the two-junction element, where we introduced for convenience $\varphi := \varphi_2$. We find

$$I(\varphi) = I_{c_1} \frac{\sin(\varphi)}{\sqrt{r^2 + 2r\cos(\varphi) + 1}}. \tag{5.33}$$

Integration of the current phase relation gives the potential energy

$$V(\varphi) = -\sqrt{E_{J_1}^2 + 2E_{J_1}E_{J_2}\cos(\varphi) + E_{J_2}^2} \tag{5.34}$$

$$= -E_{J_1} - E_{J_2} + \frac{E_{J_1}E_{J_2}}{E_{J_1} + E_{J_2}}\frac{\varphi^2}{2} - \frac{r^2 - r + 1}{r^2 + 2r + 1}\frac{E_{J_1}E_{J_2}}{E_{J_1} + E_{J_2}}\frac{\varphi^4}{24} + \dots. \tag{5.35}$$

First, it should be noted that the potential energy at the origin is simply the sum of both Josephson junction energies. This expected value can always be enforced by choosing the integration constant. Besides the constant energy offset, we recover for $r \to \infty$ the junction potential of the weaker junction. A closer look reveals that we obtain an effective Josephson energy E_J' via

$$E_J' \frac{\varphi^2}{2} := \frac{E_{J_1}E_{J_2}}{E_{J_1} + E_{J_2}}\frac{\varphi^2}{2} = \frac{\phi^2}{2(L_1 + L_2)}, \tag{5.36}$$

which shows that the inductances of the junctions can simply be added, as one would naively expect. The reduced anharmonicity given by the fourth order term shows already that junctions in series tend towards a linear inductance, as may directly be seen by the slanted current phase relation. The current phase relation and potential energy are shown in Fig. 5.6b, c.

Born-Oppenheimer approximation

The derivation of the current phase relation leads to a potential that is 2π-periodic. This result is quite confusing on first glance. We would rather expect a 4π-periodicity, since a phase drop of $\pm\pi$ can be accumulated on each of the junctions before the wave function turns into itself. Is the reduced periodicity the result of omitting the energetically unfavored solution branch? This puzzle is solved by considering the enlarged Hilbert space that emerges when the junction capacitances are included. The derivation of the Hamiltonian is discussed in more detail in the upcoming section on junction arrays. Here, we use these results in order to focus on the Born-Oppenheimer approximation.

The two-dimensional Hamiltonian of the circuit depicted in Fig. 5.5b is

$$H = \frac{1}{2}\mathbf{q}'^{T} \frac{1}{Cc_1 + Cc_2 + c_1c_2} \begin{pmatrix} C + c_2 & -C \\ -C & C + c_1 \end{pmatrix} \mathbf{q}' - E_{J_1}\cos(\varphi_1') - E_{J_2}\cos(\varphi_2'). \quad (5.37)$$

In order to simplify the analysis, we will next assume that $c_1 = c_2 =: c$. The two Josephson energies can remain independent.

Next, we can choose a new coordinate system $\boldsymbol{\phi}'' = \mathbf{S}^{-1}\boldsymbol{\phi}'$ and correspondingly $\mathbf{q}'' = \mathbf{S}^T\mathbf{q}'$, with

$$\mathbf{S} = \begin{pmatrix} \frac{1}{2} & -1 \\ \frac{1}{2} & 1 \end{pmatrix}, \qquad \mathbf{S}^{-1} = \begin{pmatrix} 1 & 1 \\ -\frac{1}{2} & \frac{1}{2} \end{pmatrix}, \quad (5.38)$$

which yields the Hamiltonian

$$H = \frac{1}{2}\mathbf{n}''^{T} \begin{pmatrix} E_C & 0 \\ 0 & E_c \end{pmatrix} \mathbf{n}'' - (E_{J_1} + E_{J_2})\cos\left(\frac{\varphi_1''}{2}\right)\cos(\varphi_2'') - (E_{J_1} - E_{J_2})\sin\left(\frac{\varphi_1''}{2}\right)\sin(\varphi_2''), \quad (5.39)$$

where we define the charging energies $E_C = 4e^2 \cdot 2/(2C + c)$ and $E_c = 4e^2/(2c)$. Since we can identify the direction φ_2'' to be always the lighter particle, we rewrite the potential as

$$V = -\operatorname{sgn}\left(E_\Sigma(\varphi_1'')\right) E_J(\varphi_1'')\cos\left(\varphi_2'' - \arctan\left(\frac{E_\Delta(\varphi_1'')}{E_\Sigma(\varphi_1'')}\right)\right), \quad (5.40)$$

with $E_J(\varphi_1'') = \sqrt{E_{J_1}^2 + 2E_{J_1}E_{J_1}\cos(\varphi_1'') + E_{J_2}^2}$, $E_\Sigma(\varphi_1'') = (E_{J_1} + E_{J_2})\cos(\varphi_1''/2)$, and $E_\Delta(\varphi_1'') = (E_{J_1} - E_{J_2})\sin(\varphi_1''/2)$. For details on the latter formula, see Sec. 5.3.2.

The Born-Oppenheimer approximation implies to determine the energy of the lighter particle in dependence of the position of the heavy particle. The energy of the light

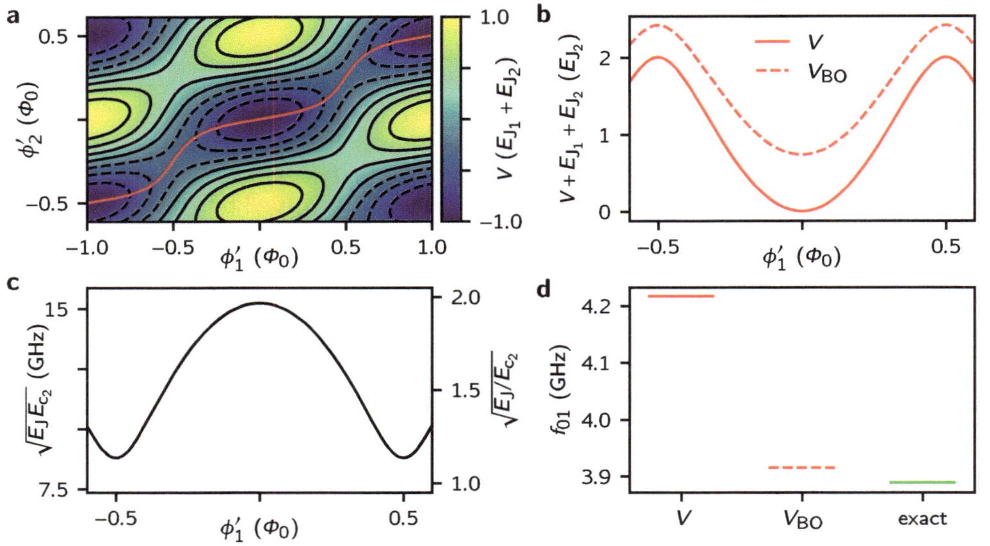

Figure 5.7: Full quantum mechanical treatment of two junctions in series. Two Josephson junctions in series with the parameters $E_{J_1} = 20\,\mathrm{GHz}$, $E_{J_2} = 10\,\mathrm{GHz}$, $C = 50\,\mathrm{fF}$, $c = 10\,\mathrm{fF}$, resulting in $E_C = 2.8\,\mathrm{GHz}$ and $E_c = 7.7\,\mathrm{GHz}$. The parameters were chosen such that the first order perturbation can be applied in good approximation. **a** Two-dimensional potential energy landscape V as a function of $\phi'_{1,2}$. The position of minimal energy is depicted by the red solid line. **b** Born-Oppenheimer potential V_{BO} experienced by the heavy particle and potential V corresponding to the red line in panel a. **c** Energy and admittance of the light particle justifying the Born-Oppenheimer and first order approximation. **d** Predicted qubit transition frequencies in comparison to the exact diagonalization using a two-dimensional charge basis.

particle becomes the potential landscape of the heavy particle. From Eq. 5.40 we see that the light particle is simply described by a charge qubit Hamiltonian centered at the position

$$\varphi''_2 = \arctan\left(\frac{E_{J_1} - E_{J_2}}{E_{J_1} + E_{J_2}}\tan\left(\frac{\varphi''_1}{2}\right)\right). \tag{5.41}$$

This dependence resembles already the classical equation, indeed insertion of φ'_i yields Eq. 5.31.

The energy of the light particle in the ground state is given by Eq. 5.30 in first order approximation. The potential of the heavy particle is therefore

$$V_{\mathrm{BO}} = -E_J(\varphi''_1) + \frac{1}{2}\sqrt{E_J(\varphi''_1)E_c} - \frac{E_c}{32}. \tag{5.42}$$

The first term gives exactly the same potential energy as derived in the previous paragraph using the current phase relation. The next two terms are essentially Lamb shift corrections of the enlarged Hilbert space. The low lying energy spectrum can now conveniently be computed in first order approximation. In Fig. 5.7 we show the potential landscape and contrast the approximated and exact qubit transition frequencies. The Born-Oppenheimer approximation yields indeed a good agreement with the exact result obtained by numerical diagonalization using a two-dimensional charge basis (cf. Eq. 5.25).

5.2.2 Josephson junction arrays

The kinetic inductance of low impedance flux qubits [211] and high impedance fluxoniums [15] is typically implemented by connecting a few up to hundreds of Josephson junctions in series. In view of the high kinetic inductance of granular aluminum, which may also be modeled as network of Josephson junctions (s. Sec. 5.2.3), it is worthwhile understanding how the inductance is formed and to get an idea of the quantum mechanical complexity.

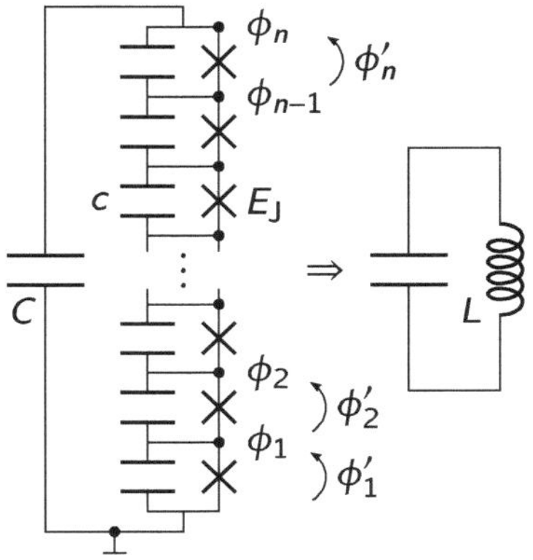

Figure 5.8: Equivalent circuit of a junction array, assuming identical junctions with Josephson energy E_J and capacitance c and that the charging of the junction islands to a common ground can be neglected. The array is shunted by the capacitance C, thereby implementing a LC-resonator shown to the right.

Junction arrays are often implemented as a stripline resonator and usually theoretically modeled as such [202, 212]. This means that each junction island has a capacitance to ground, modeling the charging energy, in addition to the capacitances to its neighboring islands. This yields a mode structure that approaches the junction plasma frequency with decreasing wavelength [202]. In contrast, junction arrays that are implemented in "free space" have essentially no capacitance to ground and only minor capacitances to distant junction islands in the array. This means that almost all modes can be found at the plasma frequency. Minimizing the capacitances is an important strategy to avoid frequency crowding, not only between neighboring quantum circuits but also within the circuit. For instance, spurious modes in the array can degrade the readout quality as well as the prediction of the qubit's dispersive shift on the resonator in the design process. In practice, this means that high impedance circuits are ideally implemented without any nearby metallic ground plane and as compact as possible, e.g. by stacking the junctions, by meandering the inductor when using kinetic inductance materials [19], or by lifting the circuit from the substrate [222].

The upcoming analysis will examine the junction array from the direction of negligible capacitance to a common ground. For simplicity, we neglect any geometric inductance between the junctions as well as spurious capacitances between the junction islands that will lift the degeneracy of all the plasma modes shifting them to lower frequencies. Even though not necessarily needed for most parts of the derivations, we will assume that the junction array capacitances as well as the Josephson energies are identical. The starting point is the circuit depicted in Fig. 5.8. Without the capacitance C shunting the junction array we would have n uncoupled charge qubits with the plasma frequency $f_{\mathrm{p}} = \sqrt{E_{\mathrm{J}}E_{\mathrm{c}}}/h$ and $E_{\mathrm{c}} = 4e^2/c$. The Lagrangian of the system is

$$\begin{aligned}\mathcal{L} = \frac{1}{2}c\dot{\phi}_1^2 + \frac{1}{2}c(\dot{\phi}_2 - \dot{\phi}_1)^2 + \cdots + \frac{1}{2}c(\dot{\phi}_n - \dot{\phi}_{n-1})^2 + \frac{1}{2}C\dot{\phi}_n^2 \\ + E_{\mathrm{J}}\cos(\varphi_1) + E_{\mathrm{J}}\cos(\varphi_2 - \varphi_1) + \cdots + E_{\mathrm{J}}\cos(\varphi_n - \varphi_{n-1}).\end{aligned} \tag{5.43}$$

In order to simplify the transition to the Hamiltonian, i.e. the calculation of the inverse of the capacitance matrix, we can transition to the primed coordinate system, as depicted in Fig. 5.8, which exploits the symmetry of the system. In this coordinate system, the Lagrangian is of the form

$$\begin{aligned}\mathcal{L} = \frac{c}{2}\dot{\boldsymbol{\phi}}'^T\begin{pmatrix}1 & & \\ & \ddots & \\ & & 1\end{pmatrix}\dot{\boldsymbol{\phi}}' + \frac{C}{2}\dot{\boldsymbol{\phi}}'^T\begin{pmatrix}1 & \dots & 1 \\ \vdots & \ddots & \vdots \\ 1 & \dots & 1\end{pmatrix}\dot{\boldsymbol{\phi}}' \\ + E_{\mathrm{J}}\cos(\varphi_1') + E_{\mathrm{J}}\cos(\varphi_2') + \cdots + E_{\mathrm{J}}\cos(\varphi_n').\end{aligned} \tag{5.44}$$

Here, we can already surmise that the interesting dynamics will happen in the space diagonal $(1 \dots 1)^T$, which sees the capacitance $nC + c$, whereas all other orthogonal directions experience the capacitance c. With this information on the eigenvalues of the capacitance matrix, we can find the inverse and arrive at the system Hamiltonian:

$$\begin{aligned}H = \frac{1}{2c}\mathbf{q}'^T\begin{pmatrix}1 & & \\ & \ddots & \\ & & 1\end{pmatrix}\mathbf{q}' + \frac{1}{2}\left(\frac{1}{n^2C + nc} - \frac{1}{nc}\right)\mathbf{q}'^T\begin{pmatrix}1 & \dots & 1 \\ \vdots & \ddots & \vdots \\ 1 & \dots & 1\end{pmatrix}\mathbf{q}' \\ + E_{\mathrm{J}}\cos(\varphi_1') + E_{\mathrm{J}}\cos(\varphi_2') + \cdots + E_{\mathrm{J}}\cos(\varphi_n').\end{aligned} \tag{5.45}$$

In the usual transmon regime, the wave function will be confined to the center of the hyper-cube. In the diagonal direction, the potential landscape amounts to

$$V(\phi', \dots, \phi') = -nE_{\mathrm{J}}\cos\left(\frac{\phi'}{\sqrt{n}}\right) = -nE_{\mathrm{J}} + \frac{1}{2}E_{\mathrm{J}}\phi'^2 - \cdots . \tag{5.46}$$

Interestingly, in contrast to the capacitance, the Josephson inductance remains unchanged. The n-fold Josephson energy is compensated by the enlarged periodicity of $\sqrt{n}\Phi_0$ in the diagonal direction. This implies an exponential decrease in n for the charge dispersion in diagonal direction, since E_{J} and $E_C = 4e^2/(nC + c)$ scale linearly with n, at least for large n (cf. Sec. 5.1.2). At first glance, the naive picture that junctions in series amounts to adding their inductances seems not correct. We are rather looking at the center of mass mode that is n-times as heavy.

In order to clarify this point even further, we can perform an orthogonal basis transformation $\mathbf{S}^T$ that maps the basis vectors of the old basis $\mathbf{e}'_i$ on the basis $\mathbf{e}''_i$. We denote the new coordinates as $\phi'' = \mathbf{S}^T \phi'$. Defining $\mathbf{e}''_1 = (1, \ldots, 1)^T$ the capacitance matrix will be diagonal in the new basis. The non-linear potential can not be diagonalized simultaneously and gives rise to a large number of non-linear coupling terms. These nonlinear interactions could be the reason for various quantum demolishing effects during the qubit readout that seem to be more detrimental in junction array fluxoniums [213] in comparison to our granular aluminum fluxoniums [203]. It should be noted that the space diagonal $\mathbf{e}''_1$ becomes increasingly orthogonal to $\mathbf{e}'_i$ with n, since $\mathbf{e}''^T_1 \cdot \mathbf{e}'_i = 1/\sqrt{n}$ vanishes asymptotically.

A coordinate transformation that yields a Hamiltonian with reduced complexity can be obtained for n being a power of two. In this case, we can use the fast Haar wavelet basis as the new coordinate system. The new basis vectors point in different corners of the hyper cube. To illustrate the pattern, we depict here the coordinate transformation for $n = 8$, which is of the form

$$\phi' = \begin{pmatrix} \frac{1}{\sqrt{8}} & \frac{1}{\sqrt{8}} & \frac{1}{\sqrt{4}} & 0 & \frac{1}{\sqrt{2}} & 0 & 0 & 0 \\ \frac{1}{\sqrt{8}} & \frac{1}{\sqrt{8}} & \frac{1}{\sqrt{4}} & 0 & -\frac{1}{\sqrt{2}} & 0 & 0 & 0 \\ \frac{1}{\sqrt{8}} & \frac{1}{\sqrt{8}} & -\frac{1}{\sqrt{4}} & 0 & 0 & \frac{1}{\sqrt{2}} & 0 & 0 \\ \frac{1}{\sqrt{8}} & \frac{1}{\sqrt{8}} & -\frac{1}{\sqrt{4}} & 0 & 0 & -\frac{1}{\sqrt{2}} & 0 & 0 \\ \frac{1}{\sqrt{8}} & -\frac{1}{\sqrt{8}} & 0 & \frac{1}{\sqrt{4}} & 0 & 0 & \frac{1}{\sqrt{2}} & 0 \\ \frac{1}{\sqrt{8}} & -\frac{1}{\sqrt{8}} & 0 & \frac{1}{\sqrt{4}} & 0 & 0 & -\frac{1}{\sqrt{2}} & 0 \\ \frac{1}{\sqrt{8}} & -\frac{1}{\sqrt{8}} & 0 & -\frac{1}{\sqrt{4}} & 0 & 0 & 0 & \frac{1}{\sqrt{2}} \\ \frac{1}{\sqrt{8}} & -\frac{1}{\sqrt{8}} & 0 & -\frac{1}{\sqrt{4}} & 0 & 0 & 0 & -\frac{1}{\sqrt{2}} \end{pmatrix} \phi'', \tag{5.47}$$

and similar for larger n. Insertion into Eq. 5.45 yields

$$H = \frac{1}{2}\mathbf{q}''^T \begin{pmatrix} \frac{1}{nC+c} & & & \\ & \frac{1}{c} & & \\ & & \ddots & \\ & & & \frac{1}{c} \end{pmatrix} \mathbf{q}'' - E_J \sum_{j=1}^{n} \cos\left(\sum_{k}^{n} \mathbf{S}_{jk} \varphi''_k \right). \tag{5.48}$$

Due to our choice of the Haar wavelet basis, one can easily see that along the axes of the new coordinate system the potential appears to be

$$\begin{aligned} V(\varphi_1'',0,\dots,0) &= nE_\mathrm{J}\cos\left(\frac{\varphi_1''}{\sqrt{n}}\right), \\ &\vdots \\ V(0,\dots,0,\varphi_{n/2}'',0,\dots,0) &= 4E_\mathrm{J}\cos\left(\frac{\varphi_{n/2}''}{\sqrt{4}}\right), \\ &\vdots \\ V(0,\dots,0,\varphi_n'') &= 2E_\mathrm{J}\cos\left(\frac{\varphi_n''}{\sqrt{2}}\right). \end{aligned} \tag{5.49}$$

However, the off-axis potential is not simply a sum of the above potentials, but entails all kinds of coupling terms to various order. The sum in the cosine functions contains always $1+\log_2(n)$ terms. Using the identity $\cos(x+y)=\cos(x)\cos(y)-\sin(x)\sin(y)$ each cosine function can be split into a sum of n products and in total n^2 products of $1+\log_2(n)$ cosine and sine functions. A basis transform that has no zeros would instead give a sum of $n2^{n-1}$ products of n cosine and sine functions. In such a basis, the diagonalization and computation of matrix elements is clearly hopeless, while in the Haar basis the identification of spurious couplings could be manageable.

A general strategy to escape the jungle of modes and couplings is (i) increasing the plasma frequency of the junctions, ideally to far beyond the resonator frequency to allow for an unhindered multi-photon readout, and (ii) increasing the admittance of the junctions so that the wave functions are well confined to the center of the hypercube. For a desired inductance, (i) and (ii) require to increase the number of junctions, which additionally yields the previously mentioned asymptotic decoupling that can also be seen in the Haar basis transformation. This pretty much pinpoints to granular aluminum and was one of the the main motivations for using it. Nevertheless, it might be worthwhile to design junction array fluxoniums and readout resonators with inductors composed of 2^n junctions.

Junction array fluxonium

By adding a junction with energy E_J' parallel to the shunt capacitance, we obtain a fluxonium Hamiltonian. The external flux dependence can be included by taking into account the finite length of the junction array capacitors. This mean that a tiny geometric inductance must be added in between the junction barriers. We will not discuss this complication and simply include the flux bias in the final Hamiltonian. One might wonder whether the fluxonium junction indeed experiences the low impedance resonance mode that we derived for the junction array. This turns out not to be the case, since the added junction term amounts to

$$E_\mathrm{J}'\cos(\varphi_1'+\dots+\varphi_n') = E_\mathrm{J}'\cos\left(\sqrt{n}\varphi_1''\right). \tag{5.50}$$

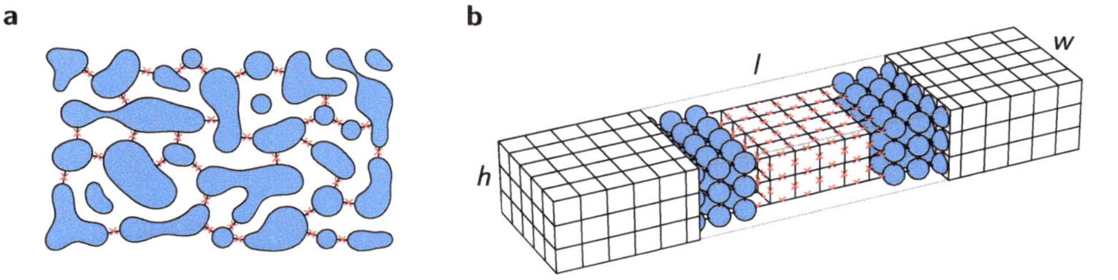

Figure 5.9: Illustration and modeling of granular aluminum. **a** Illustration of the disordered network of crystalline aluminum grains (blue) embedded in an aluminum oxide matrix. The grains are connected via tunneling barriers (red) and possibly also form constrictions. **b** Model of a granular aluminum wire assuming regularly ordered grains forming a homogeneous 3D junction network.

Introducing yet another scaling for the fluxonium degree of freedom $\phi_1''' = \sqrt{n}\phi_1''$, we obtain

$$H = \frac{1}{2}\frac{n}{nC + c}q_1'''^2 - nE_\mathrm{J}\cos\left(\frac{\phi_1''' - \Phi_\mathrm{ext}}{n}\right) - E_\mathrm{J}'\cos(\varphi_1''') + H_\perp \tag{5.51}$$

$$\approx \frac{1}{2}\frac{n}{nC + c}q_1'''^2 + \frac{1}{2}\frac{E_\mathrm{J}}{n}(\phi_1''' - \Phi_\mathrm{ext})^2 + E_\mathrm{J}'\cos(\varphi_1''') - nE_\mathrm{J} + H_\perp \tag{5.52}$$

with $H_\perp = H(\phi_1''', q_2'', \phi_2'', \ldots, q_n'', \phi_n'')$ denoting the Hamiltonian orthogonal to the qubit mode, which will disappear in the cause of the Born-Oppenheimer approximation.

5.2.3 Granular aluminum

Granular aluminum consists of crystalline aluminum grains that are embedded in an aluminum oxide matrix. The grains have a rather uniform grain size of 3 nm–4 nm [16] and form a network of Josephson junctions in the superconducting state, yielding the high kinetic inductance [17]. The disordered junction network is illustrated in Fig. 5.9a.

As long as no flux is trapped inside the granular aluminum film, the phase front of the superconducting condensate wave function will evolve in average continuously along the wire. We may therefore model the network assuming the grains are arranged regularly, as shown in Fig. 5.9b. The next approximation, assuming identical junction barriers between the grains, is certainly an oversimplification, but it allows to model the film as a junction array, since there is no phase variation across the wire [212].

Now, let l_{kin} be the kinetic inductance of a wire and a, b, and c the edge lengths of the cuboid surrounding the grain (s. Fig. 5.9b). Then, we find for the inductance of a grain as well as for the slice

$$L_{\text{gr}} = \frac{l_{\text{kin}} c}{ab}, \qquad \text{and} \qquad L_{\text{slice}} = \frac{l_{\text{kin}} c}{wh}. \tag{5.53}$$

Assuming an inductance L consisting of m slices, it must hold

$$m = \frac{L}{L_{\text{slice}}} = \sqrt{\frac{LV}{l_{\text{kin}} c^2}} = \sqrt{\frac{LV}{L_{\text{gr}} V_{\text{gr}}}}, \tag{5.54}$$

where we introduced the volume of the wire V and the volume of the grain $V_{\text{grain}} = abc$. Using the result of the previous section, the potential energy of the granular aluminum wire is

$$\begin{aligned} U &= -\frac{m\Phi_0^2}{4\pi^2 L_{\text{slice}}} \cos\left(\frac{\varphi - \varphi_{\text{ext}}}{m}\right) \\ &= -\frac{VE_{\text{gr}}}{V_{\text{gr}}} \cos\left(\sqrt{\frac{E_{\text{L}} V_{\text{gr}}}{E_{\text{gr}} V}} (\varphi - \varphi_{\text{ext}})\right). \end{aligned} \tag{5.55}$$

We neglect here the constant offset energy of all the the junctions that connect the grains within the slice. In the last step, we introduced the Josephson energy E_{gr} between the grains as well as the energy E_{L} of the inductor.

5.2.4 Granular aluminum resonator

We will now discuss a lumped element granular aluminum resonator, as implemented for instance for the readout of our granular aluminum fluxoniums [18]. The lumped resonator consists of a capacitor with charging energy E_{C} and the granular aluminum wire that we model with Eq. 5.55. The time-dependent flux can be removed as usual. Approximating the potential up to fourth order yields the Hamiltonian

$$H = \frac{E_{\text{C}}}{2} n^2 - \frac{VE_{\text{gr}}}{V_{\text{gr}}} + \frac{E_{\text{L}}}{2} \varphi^2 - \frac{E_{\text{L}}^2 V_{\text{gr}}}{VE_{\text{gr}}} \frac{\varphi^4}{24}. \tag{5.56}$$

Now, using first the approximation of Eq. 5.30 gives the nonlinear energy spectrum

$$E_n = -\frac{VE_{\text{gr}}}{V_{\text{gr}}} + \hbar\omega\left(n + \frac{1}{2}\right) - \frac{\hbar^2\omega^2}{16V} \frac{V_{\text{gr}}}{E_{\text{gr}}} \left(n^2 + n + \frac{1}{2}\right) \tag{5.57}$$

and the self-Kerr K of the resonator, here defined as $K = \hbar^2\omega^2 V_{\text{gr}}/16VE_{\text{gr}}$ [212].

As an insightful example, we can use the results of Winkel [20] to extract typical parameters of granular aluminum, i.e. the Josephson junction energy E_{gr} between the grains as well as the effective grain size c.

First of all, from the stated inductance and the size of the grAl wire, we extract $l_{\mathrm{kin}} = 11.4\,\mathrm{nHnm}$. From the specific resistivity $\rho = 1800\,\mu\Omega\mathrm{cm}$ measured at room temperature[3], we obtain a kinetic inductance resistivity ratio of

$$\alpha = 0.63\,\mathrm{nH/k\Omega}. \tag{5.58}$$

This value agrees with the predictions of the Mattis-Bardeen theory. Using the critical temperature of $T_{\mathrm{c}} \approx 2.2\,\mathrm{K}$ and the relation $\Delta = 1.76 k_{\mathrm{B}} T_{\mathrm{c}}$, a ratio of $\alpha = 0.63\,\mathrm{nH/k\Omega}$ is expected from Eq. 5.8, in agreement with the experimental findings. For higher resistivities above the dome of the critical temperature [16], where also deviations from the BCS-theory are observed [198, 214], values of $\alpha = 0.9\,\mathrm{nH/k\Omega}$ have been measured [63].

Next, from Eq. 5.57 and using the measured self-Kerr $K = 2.24\,\mathrm{MHz}$, we can extract a grain energy to volume ratio of

$$\frac{E_{\mathrm{gr}}}{V_{\mathrm{gr}}} = 1.55\,\mathrm{MHz/nm^3}, \tag{5.59}$$

and furthermore from Eq. 5.54, we obtain

$$c = \sqrt{\frac{\Phi_0^2}{4\pi^2 l_{\mathrm{kin}}} \frac{V_{\mathrm{gr}}}{E_{\mathrm{gr}}}} = 96\,\mathrm{nm}. \tag{5.60}$$

This grain size is more than one order of magnitude larger than the previously mentioned physical grain size of 3 nm–4 nm. This discrepancy has been noticed in Ref. [212] and was attributed to a potentially imprecise photon number calibration. Since the measurements of Winkel have been photon number resolving, it has instead been argued that several grains could be shunted together, forming effectively larger grain structures, as can be seen in [52].

Before going into further details, one could fear that the measured nonlinearity in Ref. [20] was superimposed by a possible contact junction between the grAl film and the shunting aluminum pads. Although this contribution was argued to be negligible, supported by measurements in magnetic field, an even more precise measurement of the nonlinearity would be desirable, which will be discussed in the next section.

[3] As was mentioned before, for a comparison with the Mattis-Bardeen formula the resistivity should be measured at the critical temperature. Fortunately, in the resistivity range that we typically use, the granular aluminum films have a residual resistance ratio (RRR) that is close to unity [17, 198].

5.2.5 Granular aluminum nonlinear interference

A precise method to measure the grAl nonlinearity is to inductively couple a resonator to a granular aluminum ring. The resonator emulates a four probe measurement of the nonlinear inductance that can be tuned with the external magnetic field, i.e. by inducing screening currents in the ring. The circuit is shown in Fig. 5.10, which was inadvertently implemented in our gradiometric fluxoniums [203]. The underlying physics is in great similarity to the superconducting nonlinear asymmetric inductive element (SNAIL) [215].

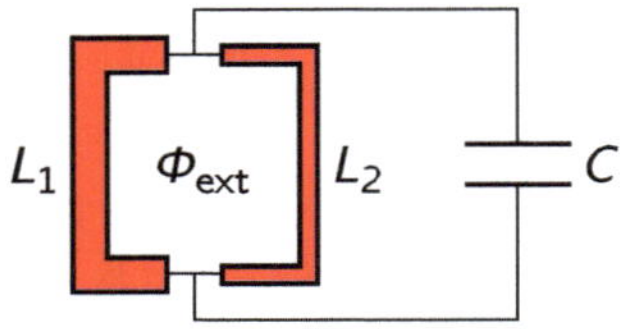

Figure 5.10: Resonator with capacitance C and field tunable granular aluminum ring inductance with inductances L_1 and L_2 enclosing the external flux Φ_{ext}.

The potential energy of both grAl wires with inductances $\mathrm{L}_{1,2}$ and volumes $V_{1,2}$ may be expressed by a Taylor expansion $U_{\mathrm{tot}} = \sum_k U_k$ following Eq. 5.55. Including the external magnetic fluxes $\varphi_{\mathrm{ext}_{1,2}}$ and a possible trapping of m flux quanta[4], it holds

$$U_k = -\frac{V_1 E_{\mathrm{gr}}}{V_{\mathrm{gr}}} \frac{\left(\sqrt{\frac{E_{\mathrm{L}_1} V_{\mathrm{gr}}}{E_{\mathrm{gr}} V_1}}(\varphi - \varphi_{\mathrm{ext}_1})\right)^{2k}}{(2k)!} - \frac{V_2 E_{\mathrm{gr}}}{V_{\mathrm{gr}}} \frac{\left(\sqrt{\frac{E_{\mathrm{L}_2} V_{\mathrm{gr}}}{E_{\mathrm{gr}} V_2}}(2\pi m - \varphi - \varphi_{\mathrm{ext}_2})\right)^{2k}}{(2k)!}. \tag{5.61}$$

Next, we parameterize the inductances $L_1 = (1 - \sin(\eta))/2 \cdot L$ and $L_2 = (1 + \sin(\eta))/2 \cdot L$ via η and the total inductance of the loop $L = L_1 + L_2$. We define the limit of linear inductors as the unperturbed Hamiltonian H_0. Introducing $\varphi' = \varphi - \varphi_{\mathrm{min}}$ shifted to the flux dependent potential minima, we find

$$H_0 = \frac{E_{\mathrm{C}}}{2} n^2 - \frac{V_1 + V_2}{V_{\mathrm{g}}} E_{\mathrm{g}} + \frac{E_{\mathrm{L}}}{2}(2\pi m - \varphi_{\mathrm{ext}})^2 + \frac{4E_{\mathrm{L}}}{2\cos^2(\eta)} \varphi'^2, \tag{5.62}$$

with $\varphi_{\mathrm{ext}} = \varphi_{\mathrm{ext}_1} + \varphi_{\mathrm{ext}_2}$. Not surprisingly, the screening current in the loop yields the flux-dependent offset energy.

Now, one can calculate the nonlinear perturbance W that arises from U_2. It can be decomposed (not recommended by hand) as a Taylor expansion in φ' with terms up to fourth order:

$$W = \sum_{l=0}^{4} c_l \varphi'^l. \tag{5.63}$$

[4]For further details on how to include the external fluxes see Sec. 5.3.1.

The prefactors are

$$
\begin{aligned}
c_0 &= -\frac{E_\mathrm{L}^2}{96}\frac{V_\mathrm{gr}}{E_\mathrm{gr}}\frac{V_2 + V_1 - 2(V_2 - V_1)\sin(\eta) + (V_1 + V_2)\sin^2(\eta)}{V_1 V_2} \cdot (2\pi m - \varphi_\mathrm{ext})^4,\\
c_1 &= -\frac{E_\mathrm{L}^2}{12}\frac{V_\mathrm{gr}}{E_\mathrm{gr}}\frac{V_2 - V_1 - (V_1 + V_2)\sin(\eta)}{V_1 V_2} \cdot (2\pi m - \varphi_\mathrm{ext})^3,\\
c_2 &= -\frac{E_\mathrm{L}^2}{4}\frac{V_\mathrm{gr}}{E_\mathrm{gr}}\frac{V_1 + V_2}{V_1 V_2} \cdot (2\pi m - \varphi_\mathrm{ext})^2,\\
c_3 &= -\frac{E_\mathrm{L}^2}{3}\frac{V_\mathrm{gr}}{E_\mathrm{gr}}\frac{V_2 - V_1 + (V_1 + V_2)\sin(\eta)}{V_1 V_2 \cos^2(\eta)} \cdot (2\pi m - \varphi_\mathrm{ext}),\\
c_4 &= -\frac{E_\mathrm{L}^2}{6}\frac{V_\mathrm{gr}}{E_\mathrm{gr}}\frac{V_2 + V_1 + 2(V_2 - V_1)\sin(\eta) + (V_1 + V_2)\sin^2(\eta)}{V_1 V_2 \cos^4(\eta)}.
\end{aligned}
\tag{5.64}
$$

With the coefficient c_4 we find again a nonlinear resonator that is independent on the magnetic field and trapped persistent currents. With increasing flux, c_3 yields a cubic perturbation and hence allows for three wave mixing processes that are utilized in SNAILs. However, since the correction on the resonator frequency first appears in second order perturbation[5], we obtain a quadratic flux dependence that is small compared to the corrections of c_2. In addition, choosing identical inductors, the third order term can completely be removed.

The main contribution comes therefore from c_2, which allows to define the flux tunable inductive energy of the resonator

$$E_\mathrm{L}^\mathrm{res} = \frac{4E_\mathrm{L}}{\cos^2(\eta)} + 2c_2. \tag{5.65}$$

The relative frequency shift of the second order term is

$$\frac{\Delta f^{(2)}}{f} = -\frac{E_\mathrm{L}\cos^2(\eta)}{16}\frac{V_\mathrm{gr}}{E_\mathrm{gr}}\frac{V_1 + V_2}{V_1 V_2}(2\pi m - \varphi_\mathrm{ext})^2. \tag{5.66}$$

Interestingly, the highest sensitivity is reached for identical inductances and with the grAl volumes as small as possible. Ideally, the grAl volumes are also chosen identical to remove the third-order nonlinearity.

[5] The third order term gives in second order perturbation an energy correction of $\frac{\Delta f_n^{(3)}}{f} = -\frac{E_\mathrm{L}\cos^2(\eta)}{4608}\sqrt{\frac{4E_\mathrm{L}E_\mathrm{C}}{\cos(\eta)^2}}\frac{V_\mathrm{gr}^2}{E_\mathrm{gr}^2}\frac{[V_2 - V_1 + (V_1 + V_2)\sin(\eta)]^2}{V_1^2 V_2^2} \cdot (30n^2 + 30n + 11) \cdot (2\pi m - \varphi_\mathrm{ext})^2$, with n being the photon number.

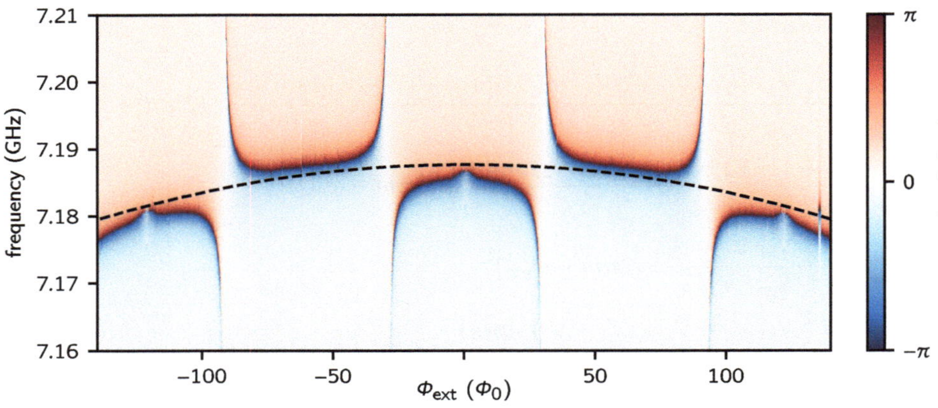

Figure 5.11: Nonlinear resonator frequency shift. Resonator phase response $\arg(S_{11})$ of the gradiometric fluxonium as a function of the external flux and of the drive frequency. The flux was calibrated with nearby non-gradiometric fluxoniums. The dashed black line shows the extracted frequency shift of $\Delta f = 6.1\,\mathrm{MHz}$ at $121\,\Phi_0$. The parameters are $C_r = 22.0\,\mathrm{fF}$, extracted from simulations and prior experiments[6], which yields $L_r + L_s = 22.3\,\mathrm{nH}$. The shared inductance $L_s \approx 2.8\,\mathrm{nH}$ was extracted from the dispersive shift by the qubit, but can also be inferred from the design. The inductance of the ring amounts to $L = 307\,\mathrm{nH}$ with a grAl volume $V = 2.97\,\mathrm{\mu m}^3$. The measurement data was taken from Ref. [203].

If both inductors are made with the same cross section such that $L_i = \alpha V_i$, the formula simplifies to

$$\frac{\Delta f^{(2)}}{f} = -\frac{1}{4}\frac{E_L}{V}\frac{V_{gr}}{E_{gr}} \cdot (2\pi m - \varphi_{ext})^2. \tag{5.67}$$

In our gradiometric fluxonium design, the resonator was implemented with an additional inductance E_{L_r} in series to the inductive ring. The corresponding inductance can simply be added to the tunable inductance of $E_L(\varphi_{ext})$, which yields a reduced sensitivity of

$$\frac{\Delta f^{(2)}}{f} = -\frac{E_{L_r}}{E_{L_r} + \frac{4E_L}{\cos^2(\eta)}}\frac{E_L\cos^2(\eta)}{16}\frac{V_{gr}}{E_{gr}}\frac{V_1+V_2}{V_1V_2} \cdot (2\pi m - \varphi_{ext})^2. \tag{5.68}$$

From the measured frequency shift that is shown in Fig. 5.11 and from the parameters of the circuit stated in the caption, we extract

$$\frac{E_{gr}}{V_{gr}} = 2.2\mathrm{MHz/nm}^3 \tag{5.69}$$

[6]In Ref. [203], the resonator inductance was inferred from the rather small shared inductance.

for a specific inductance of $l_{\mathrm{kin}} = 10.6\,\mathrm{nHnm}$. Similar to the value stated before, we obtain $c = 67\,\mathrm{nm}$.

The question remains how to explain the discrepancy between effective and physical grain size. In Ref. [90], the scaling of the effective grain size c as a function of the film resistivity is shown, estimated from the data of Ref. [212]. It shows an increasing effective grain size with the film resistivity, ranging from 10 nm to 70 nm. This could mean that an increasing disorder of the junction barriers must be taken into account. Numerical simulations of 3D junction networks seem to support this conjecture. However, solving accurately the nonlinear current phase relation in large networks is numerically challenging and time-consuming and for this reason not part of this thesis. Nevertheless, an alternative point of view is to imagine that higher resistivities appear as a smaller grAl volume, which therefore increases the nonlinearity.

5.3 The flux qubit

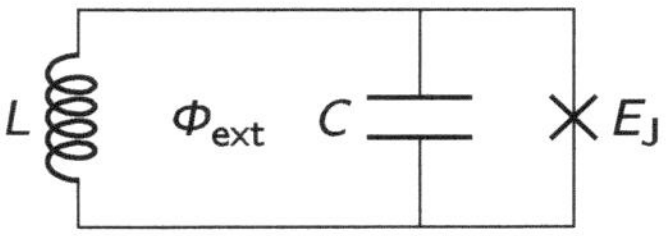

Figure 5.12: Equivalent circuit of the flux qubit. It consists of a Josephson junction with energy E_{J} that is shunted by a capacitance C and inductance L, which encloses the external flux Φ_{ext}.

In this section, a detailed analysis of flux type qubits is presented. The equivalent circuit is depicted in Fig. 5.12. The inductive shunt of the junctions prevents the qubit from being susceptible to slowly varying electric fields, which can be caused by charge noise or gate voltages in the environment. The voltage difference that drops over the Josephson junction will cause a current flow through the inductor until the electric field is compensated. This insensitivity can also be seen quantum mechanically. In its most general form, the Hamiltonian of the flux qubit reads [188, 216]

$$H = \frac{1}{2C}(q - q_{\mathrm{g}})^2 + \frac{1}{2L}(\phi - \Phi_{\mathrm{ext}})^2 - E_{\mathrm{J}} \cos\left(\frac{2\pi}{\Phi_0}\phi\right). \tag{5.70}$$

If the offset charge q_{g} is constant in time, the transformation $e^{-iq_{\mathrm{g}}\phi/\hbar} H e^{iq_{\mathrm{g}}\phi/\hbar}$ will remove the offset charge from the Hamiltonian. However, for a time-dependent offset charge, this transformation leads to additional terms in the time-dependent Schrödinger equation. In particular for ultra large inductors, charge noise could become again a non-negligible source of decoherence. Similarly, for an external flux constant in time, we may simply shift the coordinate system $\phi \longrightarrow \phi + \Phi_{\mathrm{ext}}$, which yields

$$H = \frac{1}{2C}q^2 + \frac{1}{2L}\phi^2 - E_{\mathrm{J}} \cos\left(\frac{2\pi}{\Phi_0}(\phi + \Phi_{\mathrm{ext}})\right). \tag{5.71}$$

Using the scaled coordinates introduced in Sec. 5.1.2, we obtain

$$H = \frac{1}{2}E_{\mathrm{C}}n^2 + \frac{1}{2}E_{\mathrm{L}}\varphi^2 - E_{\mathrm{J}} \cos(\varphi + \varphi_{\mathrm{ext}}), \tag{5.72}$$

where we introduced the inductive energy $E_L = \Phi_0^2/4\pi^2 L$. Furthermore, in the coordinate system of unified particle mass $m = \hbar^2$, the Hamiltonian reads

$$H = \frac{1}{2}\frac{p^2}{\hbar^2} + \frac{\hbar^2\omega^2}{2}x^2 - E_J \cos\left(\sqrt{E_C}x + \varphi_{ext}\right), \tag{5.73}$$

where the angular frequency $\omega = \sqrt{E_L E_C}/\hbar$ of the harmonic oscillator can be identified.

5.3.1 The gradiometric fluxonium

A variation of the fluxonium qubit is the gradiometric fluxonium [203, 217]. When cooled down in a magnetic field, it can be biased at half flux. In addition, it is insensitive to global field drifts. The circuit is depicted in Fig. 5.13. For simplicity, it is assumed that the inductances shunting the junction to the left and to the right are of equal size. The potential energy of the inductors is

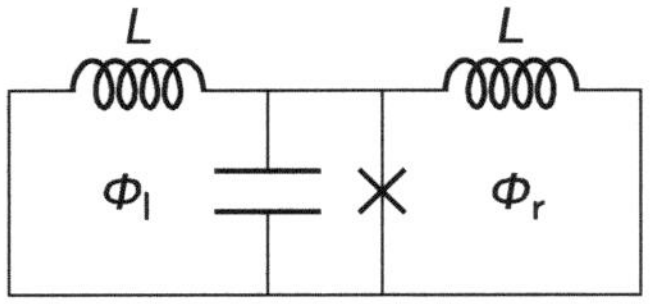

Figure 5.13: The gradiometric fluxonium with external fluxes Φ_l and Φ_r in the left and the right loop.

$$V_L = \frac{1}{2L}\left(\phi_l - \left(\frac{\Phi_s}{2} + \Phi_d\right)\right)^2 + \frac{1}{2L}\left(\phi_r - \left(\frac{\Phi_s}{2} - \Phi_d\right)\right)^2, \tag{5.74}$$

where the sum and the difference of the external fluxes $\Phi_s = \Phi_l + \Phi_r$ and $\Phi_d = (\Phi_l - \Phi_r)/2$, respectively, was introduced. The flux of the superconducting phase has to add up to integer values of flux quanta $\phi_l + \phi_r = m\Phi_0$. For clarity, $\phi = \phi_r$ is introduced. After some little algebra, one finds

$$V_L = \frac{1}{2(L/2)}\left(\phi - \left(\frac{m}{2}\Phi_0 - \Phi_d\right)\right)^2 + \frac{1}{2(2L)}(m\Phi_0 - \Phi_s)^2. \tag{5.75}$$

This shows first of all that a homogeneous external magnetic field does not enter the fluxonium Hamiltonian, but only gives rise to persistent screening currents. Only local flux sources that produce a field gradient over the sample can be used to bias the fluxonium.

Note that by blindly following the circuit quantization rules, closing the branch to the left and to right of the junction, requires introducing three flux quantization conditions, two for the left and the right loop and one for the persistent current loop. In addition, it leads to the neglect of the persistent current energy [218].

5.3.2 The SQUID fluxonium

Superconducting qubits are often implemented with a flux tunable Josephson energy by implementing the junction as a superconducting quantum interference

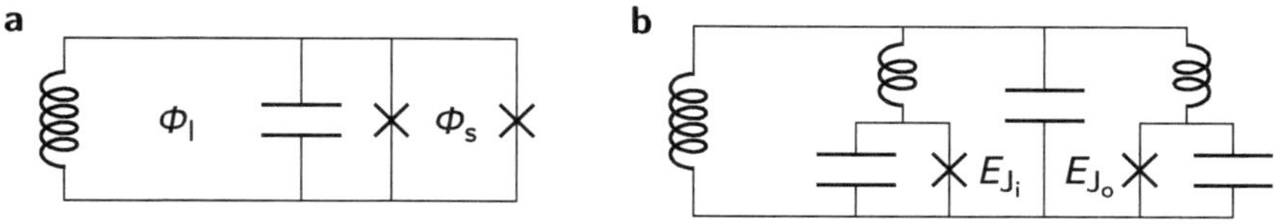

Figure 5.14: Equivalent circuit of the SQUID fluxonium. **a** Premature circuit of the fluxonium with a SQUID junction and **b** correct circuit diagram including the geometric inductance of the SQUID. The inner Josephson junction with energy E_{J_i} encloses the external flux Φ_l, whereas the outer Josephson junction with energy E_{J_o} encloses an additional flux Φ_s, which is the external flux of the SQUID.

device (SQUID). The main experimental results presented in this thesis were obtained with a SQUID fluxonium [18, 33]. For further details on the design, see Sec. 7.1.

The equivalent circuit of the SQUID fluxonium is shown in Fig. 5.14a with the SQUID formed by the two junctions. Being pedantic, the circuit diagram of the SQUID is premature. Flux biasing the SQUID in this form is not possible, since the external flux only appears in the inductor. In practice, however, the finite length of the feed lines always results in a non-negligible geometric inductance. The correct circuit is therefore the one shown in Fig. 5.14b. Its circuit quantization and its reduction to a one dimensional Hamiltonian using the Born-Oppenheimer approximation is discussed in Ref. [189].

As discussed in Sec. 5.3, for a time-independent external flux bias, the Hamiltonian can be transformed so that the flux bias appears in the potential of the Josephson junctions. The potential of the Hamiltonian is

$$V = \frac{1}{2L}\phi^2 - E_{J_i}\cos\left(\frac{2\pi}{\Phi_0}(\phi + \Phi_l)\right) - E_{J_o}\cos\left(\frac{2\pi}{\Phi_0}(\phi + \Phi_l + \Phi_s)\right). \tag{5.76}$$

Here, the inner junction with a Josephson energy E_{J_i} encloses the external flux Φ_l. The outer junction with a Josephson energy E_{J_o} encloses an additional flux Φ_s with the inner junction, which is the external flux of the SQUID. The Josephson energies in Eq. 5.76 can be rewritten as

$$\begin{aligned} &E_{J_i}\cos(\varphi + \varphi_l) + E_{J_o}\cos(\varphi + (\varphi_l + \varphi_s)) \\ &\quad = E_\Sigma(\varphi_s)\cos(\varphi + \varphi_l + \varphi_s/2) - E_\Delta(\varphi_s)\sin(\varphi + \varphi_l + \varphi_s/2) \\ &\quad = \text{sign}\left(E_\Sigma(\varphi_s)\right)\cdot\sqrt{E_\Sigma(\varphi_s)^2 + E_\Delta(\varphi_s)^2}\cdot\cos\left(\varphi + \varphi_{\text{ext}} + \arctan\left(\frac{E_\Delta(\varphi_s)}{E_\Sigma(\varphi_s)}\right)\right) \\ &\quad = E_J^{\text{eff}}(\varphi_s)\cdot\cos\left(\varphi + \varphi_{\text{ext}}^{\text{eff}}(\varphi_l, \varphi_s)\right), \end{aligned} \tag{5.77}$$

where the flux dependent energies are $E_\Sigma(\varphi_s) = (E_{J_i} + E_{J_o})\cos(\varphi_s/2)$ and $E_\Delta(\varphi_s) = (E_{J_o} - E_{J_i})\sin(\varphi_s/2)$ defining the flux tunable Josephson energy E_J^{eff} of the fluxonium. The linear external flux is given by $\varphi_{\text{ext}} = \varphi_l + \varphi_s/2$, showing that the SQUID flux

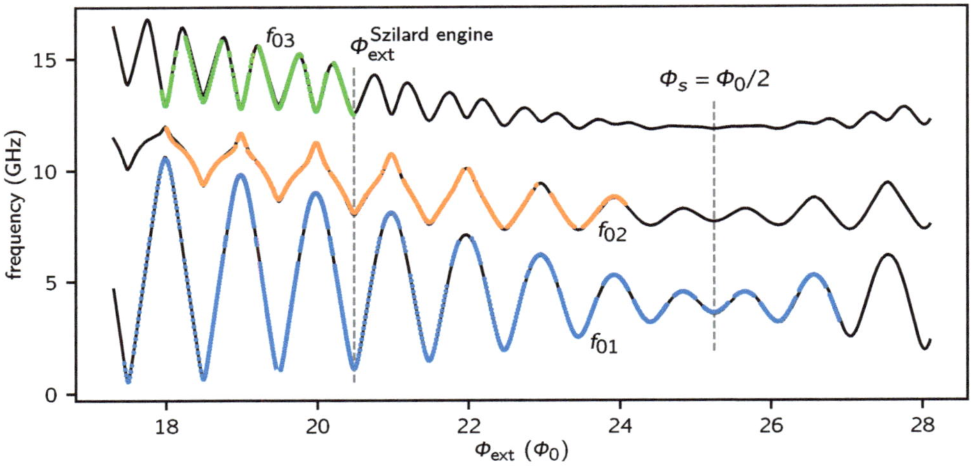

Figure 5.15: Measured and calculated energy spectrum of the granular aluminum SQUID fluxonium. At the SQUID frustration the spectrum approaches that of a harmonic oscillator. The hyperpolarization with the Szilard engine was mainly studied at the indicated flux, but also at the two neighboring flux sweet spots. The slight deviations to the theoretical model, especially at higher frequencies, are likely caused by spurious capacitances in the design. We expect the first self-resonant mode of the fluxonium superinductor in the range of 15 GHz. The extracted parameters are $L = 233.7\,\mathrm{nH}$, $C = 6.96\,\mathrm{fF}$, $E_{\mathrm{J_i}} + E_{\mathrm{J_o}} = 20.23\,\mathrm{GHz}$, $E_{\mathrm{J_o}} - E_{\mathrm{J_i}} = 0.66\,\mathrm{GHz}$, and $r = 50.0$.

contributes half to the external flux bias of the fluxonium. It should be noted that in case the two junctions are not identical, the nonlinear flux roll-off given by the inverse tangent tells apart which of the two junctions has the larger or smaller Josephson energy. In contrast, this is not possible for a charge qubit that is implemented with a SQUID. In Fig. 5.15, the measured and calculated energy spectrum of the SQUID fluxonium is shown.

In our experiment, a homogeneous magnetic field is applied to our circuit. If the areas of the enclosing loops have the ratio $r = A_\mathrm{l}/A_\mathrm{s}$, the flux in the SQUID is related to the external flux by

$$\frac{\varphi_\mathrm{s}}{2} = \frac{1}{2r+1} \cdot \varphi_\mathrm{ext}. \tag{5.78}$$

Last, it should be mentioned that for a SQUID fluxonium the condition for destructive quasiparticle interference at both Josephson junctions can no longer be met simultaneously when the SQUID is partially frustrated. The SQUID fluxonium will therefore in general not decouple from quasiparticle loss at the junction [55, 71]. The quasiparticle destructive interference condition requires the condensate to have a phase difference of π across the junction. Hence, we must have

$$\varphi_\mathrm{l} = 2\pi k + \pi \quad \wedge \quad \varphi_\mathrm{s} = 2\pi m, \qquad m, k \in \mathbb{Z}, \tag{5.79}$$

and it follows that the Josephson energies add up to $-\left(E_{\mathrm{J_i}} + E_{\mathrm{J_o}}\right)\cos(\varphi)$ resulting in a flux qubit biased at half flux, but leaving no room for tuning the Josephson energy. For example, in our quantum Szilard engine experiment, the device was operated at $\Phi_{\mathrm{ext}} = 20.48\,\Phi_0$ and designed with $r = 50.0 \pm 0.1$, therefore we have

$$\begin{aligned} \varphi_{\mathrm{s}} &= \frac{2}{2r+1}\cdot\varphi_{\mathrm{ext}} &&= 0.41\cdot 2\pi, \\ \varphi_{\mathrm{l}} &= \left(1-\frac{1}{2r+1}\right)\cdot\varphi_{\mathrm{ext}} &&= 20.28\cdot 2\pi, \\ \varphi_{\mathrm{l}}+\varphi_{\mathrm{s}} &= \left(1+\frac{1}{2r+1}\right)\cdot\varphi_{\mathrm{ext}} &&= 20.69\cdot 2\pi. \end{aligned} \tag{5.80}$$

5.3.3 Diagonalization via the harmonic oscillator basis

The wave functions of the fluxonium have to be solved numerically. For a fast convergence, it is expedient to develop the wave functions with a suitable set of basis functions. A natural choice are the eigenfunctions of the harmonic oscillator with the nonlinear Josephson potential being the perturbance [219]. The harmonic oscillator wave functions are stated in Eq. 5.27. As before, the parameter β depends on the coordinate system. In the basis of the dimensionless operators, i.e. the Hamiltonian in Eq. 5.72, we have $\beta = \sqrt[4]{E_{\mathrm{L}}/E_{\mathrm{C}}}$.

The flux and charge operators can now be expressed with the creation and annihilation operators of the harmonic oscillator. It holds

$$\frac{\phi}{\Phi_0} = \frac{\varphi}{2\pi} = \frac{1}{\sqrt{8\pi^2}}\sqrt[4]{\frac{E_{\mathrm{C}}}{E_{\mathrm{L}}}}(a^\dagger + a) = \frac{\phi_{\mathrm{ZPF}}}{\Phi_0}(a^\dagger + a), \tag{5.81}$$

$$\frac{q}{2e} = n = \frac{i}{\sqrt{2}}\sqrt[4]{\frac{E_{\mathrm{L}}}{E_{\mathrm{C}}}}(a^\dagger - a) = i\cdot\frac{q_{\mathrm{ZPF}}}{2e}(a^\dagger - a). \tag{5.82}$$

The flux and charge zero point fluctuations (ZPF) may further be simplified to yield the symmetric form

$$\frac{\phi_{\mathrm{ZPF}}}{\Phi_0} = \sqrt{\frac{Z}{4\pi R_{\mathrm{Q}}}} \qquad \frac{q_{\mathrm{ZPF}}}{2e} = \sqrt{\frac{R_{\mathrm{Q}}}{4\pi Z}}, \tag{5.83}$$

with the impedance $Z = \sqrt{L/C}$ and the resistance quantum $R_{\mathrm{Q}} = h/4e^2$. In the regime where $Z \approx R_{\mathrm{Q}}$, the flux and charge fluctuations are of similar size. In this regime, flux qubits are denoted as fluxonium qubits [15].

With respect to the harmonic oscillator basis, the Hamiltonian matrix elements are

$$\mathbf{H}_{m,n} = \langle h_m|H|h_n\rangle = \hbar\omega\left(n+\frac{1}{2}\right)\delta_{m,n} - E_{\mathrm{J}}\,\langle h_m|\cos(\varphi+\varphi_{\mathrm{ext}})|h_n\rangle\,, \tag{5.84}$$

The junction term must first be decomposed using

$$\cos(\varphi + \varphi_{\text{ext}}) = \cos(\varphi_{\text{ext}})\cos(\varphi) - \sin(\varphi_{\text{ext}})\sin(\varphi) \tag{5.85}$$

before it can be integrated analytically. It holds [220]

$$\begin{aligned}
\langle h_m|\cos(\varphi)|h_n\rangle &= \int_{\infty}^{\infty} \langle h_m|\varphi\rangle \langle\varphi|h_n\rangle \cos(\varphi)\mathrm{d}\varphi \\
&= \int_{\infty}^{\infty} \frac{\cos(by)}{\sqrt{2^{m+n}m!n!\pi}} \mathrm{H}_m(y)\mathrm{H}_n(y)e^{-y^2}\mathrm{d}y \\
&= \begin{cases} (-1)^{\frac{\tilde{m}-\tilde{n}}{2}} \dfrac{b^{\tilde{m}-\tilde{n}}}{\sqrt{2^{\tilde{m}-\tilde{n}}\tilde{m}!/\tilde{n}!}} \mathrm{L}_{\tilde{n}}^{\tilde{m}-\tilde{n}}\left(\dfrac{b^2}{2}\right) e^{-\frac{b^2}{4}}, & m+n \text{ even}, \\ 0, & m+n \text{ odd}, \end{cases}
\end{aligned} \tag{5.86}$$

where we used the substitution $y = \beta\varphi$ and defined $\tilde{m} = \max(m,n)$ and $\tilde{n} = \min(m,n)$ as well as $b = 1/\beta = \sqrt{8\pi^2}\phi_{\text{ZPF}}/\Phi_0$. Similarly, we have

$$\langle h_m|\sin(\varphi)|h_n\rangle = \begin{cases} 0, & m+n \text{ even}, \\ (-1)^{\frac{\tilde{m}-\tilde{n}-1}{2}} \dfrac{b^{\tilde{m}-\tilde{n}}}{\sqrt{2^{\tilde{m}-\tilde{n}}\tilde{m}!/\tilde{n}!}} \mathrm{L}_{\tilde{n}}^{\tilde{m}-\tilde{n}}\left(\dfrac{b^2}{2}\right) e^{-\frac{b^2}{4}}, & m+n \text{ odd}. \end{cases} \tag{5.87}$$

Numerical diagonalization of the Hamiltonian gives the eigenenergies and corresponding eigenfunctions. In Fig. 5.16a, b the wave function representation for zero and half flux is shown, while in Fig. 5.16c the eigenenergies are depicted as a function of the external flux. The qubit frequency can be tuned over a wide range and forms a well separated qubit system at half flux. In Fig. 5.16d, the speed of convergence is depicted.

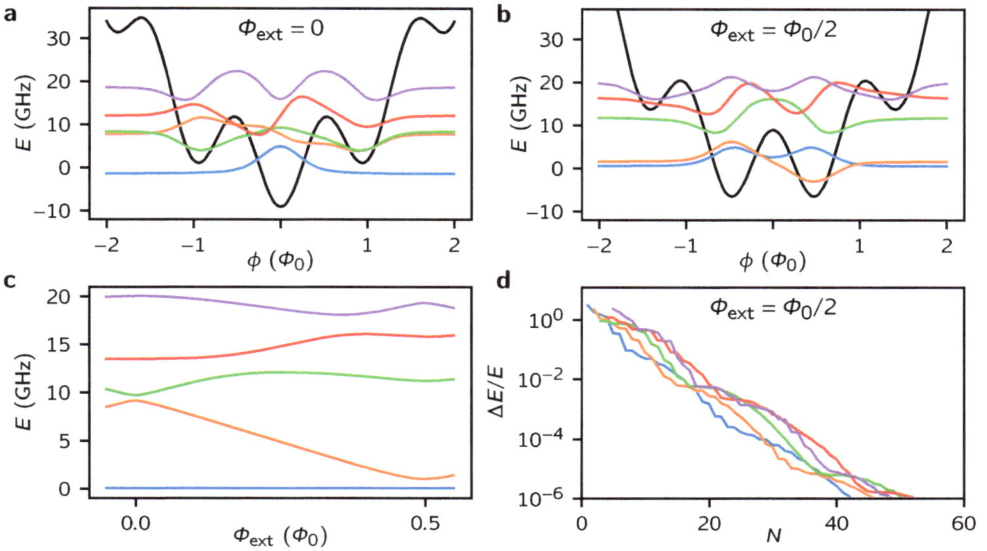

Figure 5.16: Spectrum and convergence of the fluxonium. **a** Wave function representation of the fluxonium at zero flux and **b** at half flux for $L = 300\,\text{nH}$, $C = 5\,\text{fF}$ and $E_J = 9\,\text{GHz}$. **c** Energy spectrum with respect to the ground state as a function of the external flux. **d** Relative error of the eigenenergies depending on the number of harmonic oscillator states used for the diagonalization.

5.3.4 LCAO-diagonalization

The diagonalization with the harmonic oscillator basis is only efficient as long as all energies are of comparable size. In the regime where the junction is shunted by a comparably low or ultra-high impedance resonator, other diagonalization approaches can become more efficient. Representative circuits with a comparably low impedance are the original flux qubit and the heavy fluxonium [221], while the regime of ultra-high impedance is best represented by the quasicharge[7] qubit [222], where E_L was substantially decreased by the implementation of a so-called hyperinductance. The regime of ultra-high impedance circuits can conveniently be reached using the high kinetic inductance of granular aluminum and could allow the study of high dimensional topologically protected qubits [223].

The reason why the diagonalization with the harmonic oscillator basis becomes inefficient is the fast wave function modulation induced by the cosine potential, which

[7] In my opinion, the name blochnium, which is also used is rather misleading. The ultra-high impedance is implemented in such a way as to keep the average level spacing given by $\hbar\omega = \sqrt{E_L E_C}$ in the microwave regime. The formation of Bloch bands requires instead the limit $E_L \to 0$ with E_J, E_C being finite.

requires a large number of harmonic oscillator states (s. Fig. 5.16d). Especially for higher dimensional systems, this basis is not suitable. Since the potential landscape of superconducting circuits resembles that of a molecule, it is not far to seek to use a linear combination of atomic orbitals (LCAO). The advantage is not only a smaller number of basis functions, but it also avoids the computation of high order Laguerre polynomials. However, the speed of convergence is only on credit, since the set of atomic orbital wave functions is generally overcomplete. In this case, the convergence will slow down approximately with the number of sites, and on top, the method becomes numerically unstable. When a high precision is required, one should rather look for an orthogonal wavelet basis, ideally with analytical Hamiltonian matrix elements. The LCAO method, also known as the tight-binding model, has been covered in the literature [224]. Here, we follow a slightly different approach: instead of approximating the potential with ladder operators, we derive the analytical integrals. As an instructive example, the diagonalization of an ultra-high impedance fluxonium is shown.

The eigenfunctions ψ of the Hamiltonian can be approximated by a linear superposition of atomic orbital wave functions χ_k. We therefore make the ansatz

$$\psi = \sum_k c_k \chi_k = \sum_k (a_k + ib_k)\chi_k. \tag{5.88}$$

Following the Rayleigh-Ritz method, the eigenfunctions are best approximated by ψ when the gradient of the corresponding energy

$$E(a_1, b_1, a_2, b_2, \ldots) = \frac{\langle\psi|H|\psi\rangle}{\langle\psi|\psi\rangle} \tag{5.89}$$

vanishes. This leads to the condition

$$\nabla E = \frac{1}{\langle\psi|\psi\rangle}\left(\nabla\langle\psi|H|\psi\rangle - \underbrace{\frac{\langle\psi|H|\psi\rangle}{\langle\psi|\psi\rangle}}_{E}\nabla\langle\psi|\psi\rangle\right) = 0. \tag{5.90}$$

Defining the matrices $\mathbf{H}_{m,n} = \langle\chi_m|H|\chi_n\rangle$ and $\mathbf{S}_{m,n} = \langle\chi_m|\chi_n\rangle$ leads to the generalized eigenvalue problem

$$\mathbf{Hc} = E\mathbf{Sc}. \tag{5.91}$$

For the atomic orbital states, we use the harmonic oscillator basis in each of the wells of the cosine potential. Ideally, these atomic orbitals should be positioned at the local potential minimum. Unfortunately, the positions of the minima are not equidistantly spaced but slightly shifted to the side by the harmonic potential, and the exact positions are not analytically available. However, in the regime of interest, the use of approximate positions will be quickly compensated by higher orbitals and will only affect the convergence of higher excited states. The use of equidistantly spaced atomic orbitals greatly reduces the number of matrix elements that have to be calculated.

For the atomic orbitals χ_{nl}, we therefore use the harmonic oscillator states h_n with excitation n positioned at $\varphi_l = 2\pi l - \varphi_{\text{ext}}$. The atomic orbitals functions are

$$\chi_{nl}(\varphi) = \left(\frac{\beta^2}{\pi}\right)^{1/4} \frac{1}{\sqrt{2^n n!}} \mathrm{H}_n(\beta(\varphi - \varphi_l)) e^{-\frac{\beta^2(\varphi-\varphi_l)^2}{2}}. \tag{5.92}$$

The parameter β may also be used to speed up the convergence. Here, we use $\beta = \sqrt[4]{E_\mathrm{J}/E_\mathrm{C}}$, which corresponds to the energy $\hbar\omega_\mathrm{p} = \sqrt{E_\mathrm{J}E_\mathrm{C}}$ in the junction potential, with ω_p denoting the plasma frequency.

All integrals for the calculations of the matrix elements are analytical thanks to the Taylor expansion of the Hermite polynomials

$$H_n(x+y) = \sum_{i=0}^{n} \binom{n}{i} H_i(x)(2y)^{n-i}. \tag{5.93}$$

It is useful to introduce the distance $d_{kl} = \beta(\varphi_l - \varphi_k)$ and the center $s_{kl} = \beta(\varphi_l + \varphi_k)/2$ of two atomic orbitals. For example, using the substitution $y = \beta\varphi$, the calculation of the overlap integrals simplifies to

$$\langle \chi_{mk} | \chi_{nl} \rangle = \frac{1}{\sqrt{2^{m+n} m! n! \pi}} \int_{\infty}^{\infty} H_m\left(y - s + \frac{d}{2}\right) H_n\left(y - s - \frac{d}{2}\right) e^{-(y-s)^2} e^{-\frac{d^2}{4}} \mathrm{d}y. \tag{5.94}$$

Defining $x = y - s$ and using Eq. 5.93, we obtain for the integral $I_{m,n}(d) := \langle \chi_{mk} | \chi_{nl} \rangle$ the expression

$$\begin{aligned} I_{m,n}(d) &= \frac{e^{-\frac{d^2}{4}}}{\sqrt{2^{m+n} m! n! \pi}} \int_{\infty}^{\infty} \sum_{i=0}^{m} \sum_{j=0}^{n} \binom{m}{i}\binom{n}{j} H_i(x) H_j(x) d^{m-i} (-d)^{n-j} e^{-x^2} \mathrm{d}x \\ &= \frac{e^{-\frac{d^2}{4}}}{\sqrt{2^{m+n} m! n!}} \sum_{i=0}^{\min(m,n)} 2^i i! \binom{m}{i}\binom{n}{i} d^{m-i} (-d)^{n-i}, \end{aligned} \tag{5.95}$$

where we used the orthogonality of the Hermite polynomials with respect to the Gaussian weight.

The matrix element calculations of the Hamiltonian follow the same line. First, we define the integrals

$$\mathrm{I}^{x}_{m,n}(d) = \frac{e^{-\frac{d^2}{4}}}{\sqrt{2^{m+n}m!n!\pi}} \cdot \int_{\infty}^{\infty} x H_m\left(x+\frac{d}{2}\right) H_n\left(x-\frac{d}{2}\right) e^{-x^2}\mathrm{d}x$$

$$= \sqrt{\frac{m}{2}} \cdot \mathrm{I}_{m-1,n}(d) + \sqrt{\frac{n}{2}} \cdot \mathrm{I}_{m,n-1}(d), \tag{5.96}$$

$$\mathrm{I}^{x^2}_{m,n}(d) = \frac{e^{-\frac{d^2}{4}}}{\sqrt{2^{m+n}m!n!\pi}} \cdot \int_{\infty}^{\infty} x^2 H_m\left(x+\frac{d}{2}\right) H_n\left(x-\frac{d}{2}\right) e^{-x^2}\mathrm{d}x$$

$$= \mathrm{I}^{x}_{m,n}(d) + \mathrm{I}_{m,n}(d)/2. \tag{5.97}$$

The harmonic potential can now be expressed as

$$\langle \chi_{mk}|\frac{E_\mathrm{L}\varphi^2}{2}|\chi_{nl}\rangle = \frac{E_\mathrm{L}e^{-\frac{d^2}{4}}}{\sqrt{2^{m+n}m!n!\pi}} \int_{\infty}^{\infty} H_m\left(x+\frac{d}{2}\right) \frac{(x-s)^2}{2\beta^2} H_n\left(x-\frac{d}{2}\right) e^{-x^2}\mathrm{d}x$$

$$= \frac{\hbar\omega^2}{2\omega_\mathrm{p}}\left(I^{x^2}_{m,n}(d) - 2sI^{x}_{m,n}(d) + s^2 I_{m,n}(d)\right). \tag{5.98}$$

Next, the kinetic energy has to be computed. It holds

$$\langle \chi_{mk}|-\frac{E_\mathrm{C}}{2}\frac{\partial^2}{\partial\varphi^2}|\chi_{nl}\rangle = \left(-\frac{E_\mathrm{C}}{2}\frac{\partial^2}{\partial\varphi^2}|\chi_{mk}\rangle\right)^{\dagger}|\chi_{nl}\rangle$$

$$\Leftrightarrow \quad \frac{\hbar\omega_\mathrm{p}}{4}\langle \chi_{mk}|2n+1-\left(x-\frac{d}{2}\right)^2|\chi_{nl}\rangle = \frac{\hbar\omega_\mathrm{p}}{4}\langle \chi_{mk}|2m+1-\left(x+\frac{d}{2}\right)^2|\chi_{nl}\rangle. \tag{5.99}$$

Taking the average of the left and right formula, a symmetric formula in m, n can be obtained:

$$\langle \chi_{mk}|-\frac{E_\mathrm{C}}{2}\frac{\partial^2}{\partial\varphi^2}|\chi_{nl}\rangle = \frac{\hbar\omega_\mathrm{p}}{2}\left(-I^{x^2}_{m,n}(d) + \left(m+n+1-\frac{d^2}{4}\right) I_{m,n}(d)\right). \tag{5.100}$$

Finally, the overlap with the cosine potential needs to be computed. Here, the equidistant choice of the atomic orbitals plays to our strength:

$$\langle \chi_{mk}|-E_\mathrm{J}\cos(\varphi+\varphi_\mathrm{ext})|\chi_{nl}\rangle =$$

$$= -\frac{E_\mathrm{J}e^{-\frac{d^2}{4}}}{\sqrt{2^{m+n}m!n!\pi}} \int_{\infty}^{\infty} \cos\left(\frac{x}{\beta} + \frac{2\pi l + 2\pi k}{2}\right) \mathrm{H}_m\left(x+\frac{d}{2}\right) \mathrm{H}_n\left(x-\frac{d}{2}\right) e^{-x^2}\mathrm{d}x$$

$$= -\frac{(-1)^{l+k}E_\mathrm{J}e^{-\frac{d^2}{4}}}{\sqrt{2^{m+n}m!n!\pi}} \sum_{i=0}^{m}\sum_{j=0}^{n} \binom{m}{i}\binom{n}{j} d^{m-i}(-d)^{n-j} \cdot \int_{\infty}^{\infty} \cos(bx) H_i(x) H_j(x) e^{-x^2}\mathrm{d}x$$

$$= -(-1)^{k+l} E_\mathrm{J}\, \mathrm{I}^{\cos}_{m,n}(d,b), \tag{5.101}$$

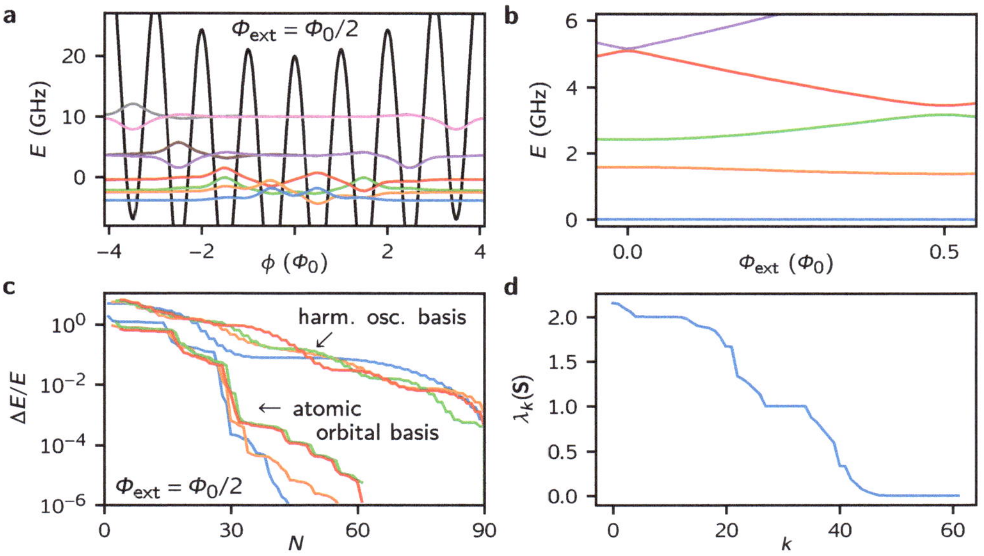

Figure 5.17: Spectrum and convergence of an ultra-high impedance fluxonium. a Wave representation of the ultra-high impedance fluxonium at half flux for $L = 3000$ nH, $C = 2$ fF and $E_J = 20$ GHz. For the diagonalization, the ground state orbitals in the lowest 8 wells and additional 54 orbitals in the lowest 6 wells are used. **b** Energy spectrum with respect to the ground state as a function of the external flux. **c** Comparison of the relative error of the eigenenergies in dependence of the number of basis functions used for the diagonalization. **d** Eigenvalues of the **S** matrix sorted in descending order, showing that out of the 62 orbitals only about 40 are linearly independent.

where we define again $b = 1/\beta$. The solution of the remaining integral was given in the previous section. For non equidistantly spaced orbitals, one also needs the integral $\mathrm{I}^{\sin}_{m,n}(d,b)$, for which simply the cosine in $\mathrm{I}^{\cos}_{m,n}(d,b)$ has to be replaced with the sine function. All together, the Hamiltonian matrix elements read

$$\begin{aligned}\mathbf{H}_{mk,nl} = {} & \frac{\hbar}{2}\left(\omega_\mathrm{p}\left(m+n+1-\frac{d_{kl}^2}{4}\right)+\frac{\omega^2}{\omega_\mathrm{p}}s_{kl}^2\right)\mathrm{I}_{m,n}(d_{kl}) \\ & + \hbar\frac{\omega^2}{\omega_\mathrm{p}}s_{kl}\,\mathrm{I}^{x}_{m,n}(d_{kl}) + \hbar\frac{\omega^2-\omega_\mathrm{p}^2}{2\omega_\mathrm{p}}\,\mathrm{I}^{x^2}_{m,n}(d) - (-1)^{k+l}E_\mathrm{J}\,\mathrm{I}^{\cos}_{m,n}(b,d_{kl}).\end{aligned} \tag{5.102}$$

The diagonalization of an ultra-high impedance fluxonium is shown in Fig. 5.17a. In Fig. 5.17b, it can be seen that the energy dispersion of the ultra-high impedance fluxonium becomes flat with respect to the applied external flux [222]. Therefore, the ultra-high impedance fluxonium plays a similar role in relation to the flux qubit as the transmon does for the charge qubit. The speed of convergence of the LCAO-method is greatly improved compared to the diagonalization using the harmonic oscillator states, as can be seen in Fig. 5.17c. The fact that mainly the orbitals of the six lowest sites have been used shows that roughly five states per site are required to obtain a good convergence. This will also be discussed in more detail in the next section. For more than five orbitals per site, the convergence slows down and becomes numerical unstable for more that ten orbitals per site. This can be explained by Fig. 5.17d showing the eigenvalues of the **S** matrix in descending order. From the 62 orbitals used for the diagonalization only about 40 states are linearly independent.

5.3.5 Fluxonium qubit approximation formula

In this section, the LCAO-method will be used to derive an approximation for the fluxonium qubit frequency biased at half flux. The approximation using the two ground state orbitals in each of the wells is not very precise due to the shape of the Gaussian wave function and approximation of the local minima. Nevertheless, the formula provides a good estimate in the regime $\sqrt{E_J/E_C} < 1$, where the charge dispersion formula becomes unprecise (cf. Fig. 5.4).

Let $|\chi_0\rangle = |\mathrm{l}\rangle$ and $|\chi_1\rangle = |\mathrm{r}\rangle$ denote the Gaussian wave functions in the left and the right well, respectively. Close to half flux their center positions are $\varphi_\mathrm{l} = -\delta/2 - (\varphi_\mathrm{ext} - \pi)$ and $\varphi_\mathrm{r} = \delta/2 - (\varphi_\mathrm{ext} - \pi)$ in first order in the external flux. Here, we introduced the distance δ between the states, which is kept as free parameter that can in principle be used to minimize the ground state energy. For instance, δ may be chosen to match the exact minima of the wells at half flux. This, however, requires to solve the transcendental equation $\delta/2 = E_J/E_L \sin(\delta/2)$ numerically. In first order approximation, this equation yields $\delta \approx 2\pi/(1 + E_L/E_J)$, valid for $E_J \gg E_L$. With only the ground states present $m = n = 0$, the calculations of the matrix elements following Eq. 5.102 are greatly simplified and amount to

$$\mathbf{S}_{ij} = e^{-\frac{d_{ij}^2}{4}} \quad \rightarrow \quad \mathbf{S} = \begin{pmatrix} 1 & e^{-\frac{\beta^2\delta^2}{4}} \\ e^{-\frac{\beta^2\delta^2}{4}} & 1 \end{pmatrix}, \tag{5.103}$$

$$\mathbf{H}_{ij} = \left[\frac{E_C\beta^2}{2}\left(\frac{1}{2} - \frac{d_{ij}^2}{4}\right) + \frac{E_L}{2\beta^2}\left(\frac{1}{2} + s_{ij}^2\right) - E_J e^{-\frac{b^2}{4}} \cos\left(b s_{ij} + \varphi_\mathrm{ext}\right) \right] \cdot e^{-\frac{d_{ij}^2}{4}}. \tag{5.104}$$

The approximate eigenenergies are the eigenvalues of the matrix $\mathbf{G} = \mathbf{S}^{-1}\mathbf{H}$. We have

$$\mathbf{G} = \frac{1}{1 - e^{-\frac{\beta^2\delta^2}{2}}} \cdot \begin{pmatrix} \mathbf{H}_{00} - \mathbf{H}_{10} e^{-\frac{\beta^2\delta^2}{4}} & \mathbf{H}_{01} - \mathbf{H}_{11} e^{-\frac{\beta^2\delta^2}{4}} \\ \mathbf{H}_{10} - \mathbf{H}_{00} e^{-\frac{\beta^2\delta^2}{4}} & \mathbf{H}_{11} - \mathbf{H}_{01} e^{-\frac{\beta^2\delta^2}{4}} \end{pmatrix}. \tag{5.105}$$

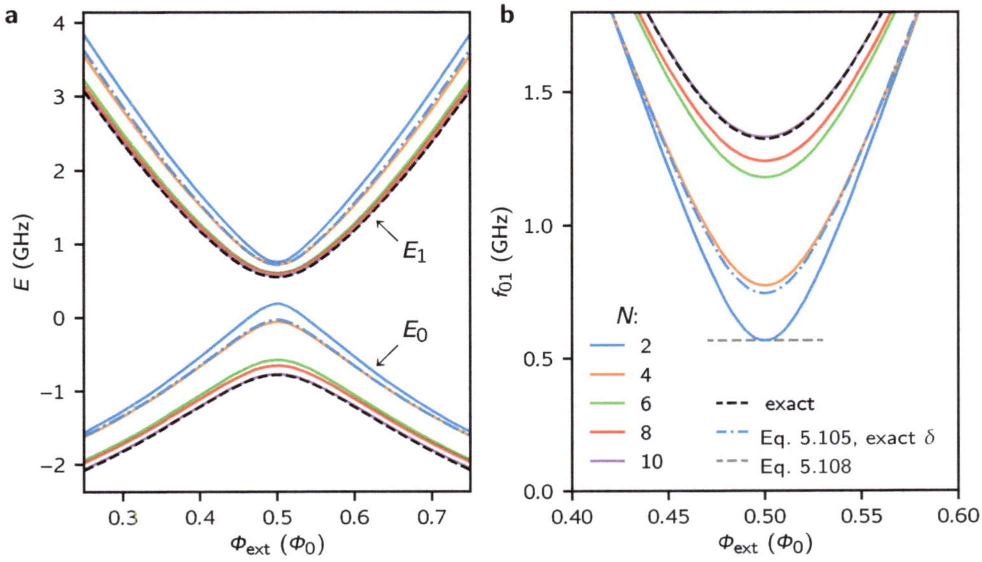

Figure 5.18: Convergence of the LCAO method for a typical fluxonium. **a** Approximation of the energies and **b** of the qubit frequency for an increasing number of orbitals N in total. For the dashed blue line, the exact distance of the potential minima at half flux is used for δ and $\beta = \sqrt{E_\mathrm{J}/E_\mathrm{C}}$. The parameters are $L = 300\,\mathrm{nH}$, $C = 5\,\mathrm{fF}$ and $E_\mathrm{J} = 15\,\mathrm{GHz}$.

Next, we will derive an approximation formula exactly at half flux where the symmetry $\mathbf{G}_{00} = \mathbf{G}_{11}$ can be exploited. The eigenenergies $E_{0,1}$ of the fluxonium and the fluxonium qubit frequency $f_{01} = (E_1 - E_0)/h$ are then given by

$$E_{0,1} = \mathbf{G}_{00} \pm |\mathbf{G}_{01}| \tag{5.106}$$

$$hf_{01} = 2|\mathbf{G}_{01}| = \left|\frac{\beta^4\delta^2}{4}E_\mathrm{C} - 4E_\mathrm{J}e^{-\frac{1}{4\beta^2}} + \frac{\delta^2}{4}E_\mathrm{L}\right| \cdot \frac{e^{-\frac{\beta^2\delta^2}{4}}}{1 - e^{-\frac{\beta^2\delta^2}{4}}} \tag{5.107}$$

$$\approx \left(\left(\pi^2 - 4e^{-\frac{1}{4}\sqrt{E_\mathrm{C}/E_\mathrm{J}}}\right)E_\mathrm{J} + \pi^2 E_\mathrm{L}\right) \cdot \frac{e^{-\pi^2\sqrt{E_\mathrm{J}/E_\mathrm{C}}}}{1 - e^{-\pi^2\sqrt{E_\mathrm{J}/E_\mathrm{C}}}}. \tag{5.108}$$

In the last step, the approximations $\delta = 2\pi$ and $\beta = \sqrt[4]{E_\mathrm{J}/E_\mathrm{C}}$ were inserted. The denominator may be approximated as well, since the overlap of the left and the right wave function is usually quite small. We keep it here for completeness.

In order to compare this approximation to the charge dispersion of the charge qubit, we rewrite $E_J = hf_p\sqrt{E_J/E_C}$, with f_p being the plasma frequency, and consider the limit $\sqrt{E_J/E_C} > 1$, assuming as well $E_J \gg E_L$. This gives the limit behavior

$$f_{01}/f_p \approx (\pi^2 - 4)\sqrt{\frac{E_J}{E_C}}e^{-\pi^2\sqrt{\frac{E_J}{E_C}}}, \tag{5.109}$$

which approaches faster to zero than the charge dispersion formula (cf. Eq. 5.24), reading for the first band

$$f_{01}/f_p \approx \frac{16}{\sqrt{\pi}}\sqrt[4]{\frac{E_J}{E_C}}e^{-8\sqrt{\frac{E_J}{E_C}}}. \tag{5.110}$$

The reason for this distinct limit behavior is that the orbitals are too strongly confined by approximating the cosine with a quadratic potential. Due to the Gaussian wave functions in the ground state, this approximation enters exponentially in the overlap integrals.

In Fig. 5.18, the diagonalization of a typical fluxonium is shown. Indeed, the approximation of the ground state is not very accurate. The comparison of the orange and dashed blue line shows that the first excited state orbitals are needed to shift the wave functions to the exact potential minima. The second excited state orbital, shifted by the third, is needed to adjust the frequency to the cosine potential. An excellent agreement with the exact energies is achieved with five orbitals in each well. For typical fluxonium qubits, the gain in computational complexity is only minor. Around 20 states are needed for the harmonic oscillator basis (cf. Fig. 5.16) compared to the 10 states for the atomic orbital basis. However, if the orbital positions and frequencies are optimized beforehand, the choice of selected orbitals, e.g. the ground state and the fifth state, may already give a satisfying approximation of the lower energy spectrum. For the diagonalization of multidimensional circuits, this gives of course a huge computational speedup, especially considering that the computational complexity of matrix diagonalization grows with $O(N^3)$.

6 Dispersive readout

A key ingredient for the successful implementation of quantum error correction codes but also for the quantum Szilard engine is the realization of a textbook quantum mechanical measurement. The ideas of realizing such ideal measurements have been put forward in the field of gravitational wave detectors [225] and resulted in the concept of quantum non-demolishing (QND) measurements. While in some experimental platforms, such as trapped ions, the QND readout is naturally available and allows for the observation of quantum jumps [166, 226, 227], sophisticated interactions in systems with harmonic potentials have to be engineered to repeatedly observe single photons without destroying them [228, 229]. In a similar fashion, the QND readout of superconducting devices is typically implemented [34, 35] and enabled us to read out the granular aluminum fluxonium using large photon numbers [33].

Since the measurement process is still far from understood, e.g. the cause and timing of the wave function collapse, one usually imagines that the system Hamiltonian H_s is coupled via the interaction Hamiltonian H_{int} to the outside world. The outside world somehow implements the measurement apparatus, which performs a textbook measurements with the operator O. The idea of the QND measurement is now as follows [230]: In order to obtain a precise measurement result, it should be possible to measure the system repetitively. Once the system got projected, it should not evolve in time so that it can be measured again. This can be achieved when the measurement operator commutes with the system Hamiltonian, i.e. $[H_s, O] = 0$. Similarly, the system should also not act back on the measurement apparatus and modify the measurement signal, which requires $[H_{int}, O] = 0$. An ideal Hamiltonian for QND measurements of a qubit would comprise a longitudinal interaction of the form

$$H = \hbar\omega_r a^\dagger a + \frac{\hbar\omega_q}{2}\sigma_z + \frac{\hbar\chi}{2}a^\dagger a\sigma_z, \tag{6.1}$$

where ω_r and ω_q denote the resonator and qubit frequencies, respectively. The interaction between the qubit and the resonator yields a mutual dispersive frequency shift that is encoded in χ. The detection of resonator photons leads in first place to a fluctuating photon number giving rise to qubit dephasing. However, once the environment has "decided" on the qubit state and absorbed all resonator photons with a specific frequency, the qubit is projected in one of its eigenstates. The energy difference before and after the projection must either be provided by the measurement apparatus [231–233] or by the qubit environment [7, 86] (s. Sec. 3.6.3).

In case it is meaningful to assign the excitation numbers n_r and n_a for the resonator and the artificial atom, respectively, to the energy spectrum, the resonator transition frequencies ω_r can be extracted via

$$\hbar\omega_{\mathrm{r},|n_\mathrm{a}\rangle}(n_\mathrm{r}) = E_{|n_\mathrm{r},n_\mathrm{a}\rangle} - E_{|n_\mathrm{r}-1,n_\mathrm{a}\rangle}. \tag{6.2}$$

Similarly, the transitions frequencies ω_{ij} of the artificial atom can be obtained via

$$\hbar\omega_{ij}(n_\mathrm{r}) = E_{|n_\mathrm{r},j\rangle} - E_{|n_\mathrm{r},i\rangle} \tag{6.3}$$

in dependence of the number of photons in the resonator. The dispersive frequency shift χ can conveniently be defined as a relative quantity between two different atomic states. Hence, we define

$$\begin{aligned}\hbar\chi_{ij}(n_\mathrm{r}) &= \hbar\omega_{\mathrm{r},|j\rangle}(n_\mathrm{r}) - \hbar\omega_{\mathrm{r},|i\rangle}(n_\mathrm{r}) &(6.4)\\ &= \hbar\omega_{ij}(n_\mathrm{r}) - \hbar\omega_{ij}(n_\mathrm{r}-1). &(6.5)\end{aligned}$$

Often, the shift with respect to the ground state $i = 0$ is considered.

6.1 Jaynes-Cummings model

A convenient way to obtain approximately the longitudinal interaction in Eq. 6.1 is the dispersive readout scheme [34, 35, 234]. Here, a transverse coupling with coupling strength g is paired with a large detuning $\delta \gg g$ between the resonator and the qubit frequencies. Although analytical solutions for the quantum Rabi model have been found in recent years [235], the concept of the dispersive shift can be well understood with the Jaynes-Cummings model employing the rotating-wave approximation:

$$H = \hbar\omega_\mathrm{r}\left(n_\mathrm{r} + \frac{1}{2}\right) + \frac{\hbar\omega_\mathrm{q}}{2}\sigma_z + \hbar g(a^\dagger\sigma^- + \sigma^+ a). \tag{6.6}$$

As will be shown in Fig. 6.5a, the rotating-wave approximation (RWA) is only valid as long as the coupling between resonator and qubit states is small, i.e. $g\sqrt{n_\mathrm{r}} \ll \omega_\mathrm{r}, \omega_\mathrm{q}$. Thus, the RWA only holds for low photons numbers and we can expect avoided level crossings in power. Indeed, we see in Fig. 6.5a that the dispersive shift undergoes an oscillating behavior as a function of the photon number.

Continuing with the Jaynes-Cummings model and assuming a negative detuning $\delta = \omega_\mathrm{q} - \omega_\mathrm{r} < 0$, the energies are given by

$$E_{|n_\mathrm{r},\mathrm{g}\rangle} = \hbar\omega_\mathrm{r} n_\mathrm{r} + \frac{\hbar}{2}\sqrt{\delta^2 + 4g^2 n_\mathrm{r}}, \tag{6.7}$$

$$E_{|n_\mathrm{r},\mathrm{e}\rangle} = \hbar\omega_\mathrm{r}(n_\mathrm{r}+1) - \frac{\hbar}{2}\sqrt{\delta^2 + 4g^2(n_\mathrm{r}+1)}, \tag{6.8}$$

where we should keep in mind that the corresponding wave functions are a mixture of qubit and resonator states. For clarity, the notation $|\mathrm{g}\rangle$ for $n_\mathrm{a} = 0$ and $|\mathrm{e}\rangle$ for $n_\mathrm{a} = 1$ is used here.

For moderate photon numbers, it holds

$$\hbar\omega_{\mathrm{r},|\mathrm{g}\rangle} = \hbar\omega_\mathrm{r} + \frac{\hbar}{2}\left(\delta - \sqrt{\delta^2 - 4g^2}\right) - \hbar g\left(\frac{g}{\sqrt{\delta^2 - 4g^2}} - \frac{g}{\delta}\right) n_\mathrm{r} + \dots \tag{6.9}$$

$$\approx \hbar\omega_\mathrm{r} + \frac{\hbar g^2}{\delta} + \frac{2\hbar g^4}{\delta^3} n_\mathrm{r} + \dots, \tag{6.10}$$

$$\hbar\omega_{\mathrm{r},|\mathrm{e}\rangle} = \hbar\omega_\mathrm{r} - \frac{\hbar}{2}\left(\sqrt{\delta^2 + 4g^2} - \delta\right) + \hbar g\left(\frac{g}{\delta} - \frac{g}{\sqrt{\delta^2 + 4g^2}}\right) n_\mathrm{r} + \dots \tag{6.11}$$

$$\approx \hbar\omega_\mathrm{r} - \frac{\hbar g^2}{\delta} - \frac{2\hbar g^4}{\delta^3} n_\mathrm{r} + \dots, \tag{6.12}$$

with the approximations valid for $\delta \gg g$. For large photon numbers, the dispersive shift vanishes slowly according to

$$\hbar\omega_{\mathrm{r},|\mathrm{g}\rangle} = \hbar\omega_\mathrm{r} + \frac{\hbar g}{2\sqrt{n_\mathrm{r}}} + \dots, \qquad \hbar\omega_{\mathrm{r},|\mathrm{e}\rangle} = \hbar\omega_\mathrm{r} - \frac{\hbar g}{2\sqrt{n_\mathrm{r}}} + \dots. \tag{6.13}$$

Since the signal-to-noise ratio (SNR) of the readout increases as $\sqrt{n_\mathrm{r}}$, the integration time to measure a certain number of signal photons has a lower limit. However, this bound is very low. In practice, and if the qubit can handle that many photons, the resonator coupling to the transmission line marks the speed limit for the state discrimination of the qubit [236].

6.2 Fluxonium with readout resonator

Nowadays, the standard way to read out superconducting artificial atoms is the dispersive readout scheme, as introduced in the previous section. For this, a readout mode, which can be implemented either as a cavity or as a lumped element resonator, is weakly coupled inductively or capacitively to the artificial atom. In Fig. 6.1, the equivalent circuit diagram of the fluxonium together with its readout resonator is shown for both coupling types. In the following, the system Hamiltonian is derived for both coupling types. Furthermore, it will be shown that both coupling strategies can yield the same energy spectrum.

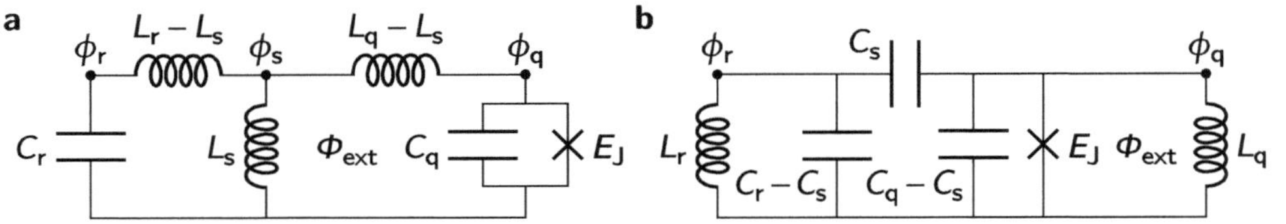

Figure 6.1: Dispersive fluxonium readout. Equivalent circuit diagram of the readout resonator being **a** inductively or **b** capacitively coupled to a fluxonium.

Starting with the inductive coupling, the Lagrangian reads:

$$\mathcal{L} = \frac{1}{2} C_r \dot{\phi}_r^2 + \frac{1}{2} C_q \dot{\phi}_q^2 + E_J \cos\left(\frac{\Phi_0}{2\pi}\phi_q\right) - \frac{(\phi_r - \phi_s - \Phi_{ext,r})^2}{2(L_r - L_s)} - \frac{(\phi_s - \Phi_{ext,s})^2}{2L_s} - \frac{(\phi_q - \phi_s - \Phi_{ext,q})^2}{2(L_q - L_s)}. \quad (6.14)$$

Here, C_r and C_q are the capacitances, L_r and L_q the inductances, and ϕ_r and ϕ_q the fluxes of the resonator and the fluxonium with Josephson energy E_J, respectively. The shared inductance and corresponding flux are denoted by L_s and ϕ_s.

It is worthwhile to introduce the gauge invariant fluxes to speed up the upcoming analysis. For example, we can write

$$\phi_r - \phi_s - \Phi_{ext,r} = (\phi_r - (\Phi_{ext,r} + \Phi_{ext,s})) - (\phi_s - \Phi_{ext,s}) = \tilde{\phi}_r - \tilde{\phi}_s. \quad (6.15)$$

This means that the external flux has to be tracked through the inductive branches in the same way as the superconducting phase.

The equation of motion has a constraint that needs to be removed. We have

$$\begin{aligned} \frac{d}{dt}\frac{\partial \mathcal{L}}{\partial \dot{\phi}_s} &= \frac{\partial \mathcal{L}}{\partial \phi_s} = \frac{\partial \mathcal{L}}{\partial \tilde{\phi}_s} \\ \Leftrightarrow \qquad 0 &= \frac{\tilde{\phi}_r - \tilde{\phi}_s}{L_r - L_s} + \frac{\tilde{\phi}_q - \tilde{\phi}_s}{L_q - L_s} - \frac{\tilde{\phi}_s}{L_s} \\ \Leftrightarrow \qquad \tilde{\phi}_s &= \frac{(L_r L_s - L_s^2)\tilde{\phi}_r + (L_q L_s - L_s^2)\tilde{\phi}_q}{L_r L_q - L_s^2}. \end{aligned} \quad (6.16)$$

Insertion of $\tilde{\phi}_s$ into Eq. 6.14 yields an unconstrained Lagrangian, which can be Legendre-transformed to give the system Hamiltonian

$$H = \frac{1}{2}\mathbf{q}^T \begin{pmatrix} \frac{1}{C_r} & 0 \\ 0 & \frac{1}{C_q} \end{pmatrix} \mathbf{q} + \frac{1}{2}\frac{1}{L_r L_q - L_s^2} \boldsymbol{\phi}^T \begin{pmatrix} L_q & -L_s \\ -L_s & L_r \end{pmatrix} \boldsymbol{\phi} - E_J \cos\left(\frac{\Phi_0}{2\pi}\phi_q\right), \quad (6.17)$$

with $\boldsymbol{\phi} = (\phi_\mathrm{r} - (\Phi_\mathrm{ext,r} + \Phi_\mathrm{ext,s}), \phi_\mathrm{q} - (\Phi_\mathrm{ext,q} + \Phi_\mathrm{ext,s}))^T$ and $\mathbf{q} = (q_\mathrm{r}, q_\mathrm{q})^T$. As usual, the time-independent external flux can be removed in the resonator, while for the fluxonium we define $\Phi_\mathrm{ext} = \Phi_\mathrm{ext,q} + \Phi_\mathrm{ext,s}$, which is the total flux enclosed by the junction. Moving to the natural basis, we can define the capacitive and inductive energies and matrices via

$$H = \frac{1}{2}\mathbf{n}^T \begin{pmatrix} E_{\mathrm{C_r}} & 0 \\ 0 & E_{\mathrm{C_q}} \end{pmatrix} \mathbf{n} + \frac{1}{2}\boldsymbol{\varphi}^T \begin{pmatrix} E_{\mathrm{L_r}} & -E_{\mathrm{L_s}} \\ -E_{\mathrm{L_s}} & E_{\mathrm{L_q}} \end{pmatrix} \boldsymbol{\varphi} - E_\mathrm{J}\cos(\varphi_\mathrm{q}) \tag{6.18}$$

$$= \frac{1}{2}\mathbf{n}^T \mathbf{E}_\mathrm{C}\mathbf{n} + \frac{1}{2}\boldsymbol{\varphi}^T\mathbf{E}_\mathrm{L}\boldsymbol{\varphi} - E_\mathrm{J}\cos(\varphi_\mathrm{q})\,. \tag{6.19}$$

In contrast, when the resonator is capacitively coupled to the fluxonium, there is no constraint variable, and the Hamiltonian can directly be stated:

$$H = \frac{1}{2}\frac{1}{C_\mathrm{r}C_\mathrm{q} - C_\mathrm{s}^2}\mathbf{q}^T \begin{pmatrix} C_\mathrm{q} & C_\mathrm{s} \\ C_\mathrm{s} & C_\mathrm{r} \end{pmatrix} \mathbf{q} + \frac{1}{2}\boldsymbol{\phi}^T \begin{pmatrix} \frac{1}{L_\mathrm{r}} & 0 \\ 0 & \frac{1}{L_\mathrm{q}} \end{pmatrix} \boldsymbol{\phi} - E_\mathrm{J}\cos\left(\frac{\Phi_0}{2\pi}\phi_\mathrm{q}\right) \tag{6.20}$$

$$= \frac{1}{2}\mathbf{n}^T \mathbf{E}_\mathrm{C}\mathbf{n} + \frac{1}{2}\boldsymbol{\varphi}^T\mathbf{E}_\mathrm{L}\boldsymbol{\varphi} - E_\mathrm{J}\cos(\varphi_\mathrm{q})\,. \tag{6.21}$$

There are two main approaches to diagonalize the Hamiltonian numerically. First, we show the simpler version, which is computationally less demanding, especially when high photon numbers in the resonator are considered. Here, one simply uses the product states

$$\langle \varphi_r\varphi_q | h_m h_n \rangle = h_m(\varphi_\mathrm{r})h_n(\varphi_\mathrm{q}) \tag{6.22}$$

with the harmonic oscillator stated as defined in Eq. 5.27 with $\beta_\mathrm{r} = \sqrt[4]{E_{\mathrm{L_r}}/E_{\mathrm{C_r}}}$ and $\beta_\mathrm{q} = \sqrt[4]{E_{\mathrm{L_q}}/E_{\mathrm{C_q}}}$. The calculation of the matrix elements $\mathbf{H}_{mn,kl} = \langle h_m h_n | H | h_k h_l \rangle$ can be separated into fluxonium and resonator terms, except for the coupling, which can be expressed as

$$-E_{\mathrm{L_s}}\varphi_\mathrm{r}\varphi_\mathrm{q} = -E_{\mathrm{L_s}}\varphi_\mathrm{r}^\mathrm{ZPF}\varphi_\mathrm{q}^\mathrm{ZPF}(a_\mathrm{r}^\dagger + a_\mathrm{r})(a_\mathrm{q}^\dagger + a_\mathrm{q}), \qquad \text{for inductive,}$$

$$E_{\mathrm{C_s}}q_\mathrm{r}q_\mathrm{q} = -E_{\mathrm{C_s}}q_\mathrm{r}^\mathrm{ZPF}q_\mathrm{q}^\mathrm{ZPF}(a_\mathrm{r}^\dagger - a_\mathrm{r})(a_\mathrm{q}^\dagger - a_\mathrm{q}), \qquad \text{for capacitive}$$

coupling. This means that the coupling is almost identical except that the rotating terms $a_\mathrm{r}^\dagger a_\mathrm{q}$ and $a_\mathrm{q}^\dagger a_\mathrm{r}$ and counter rotating terms $a_\mathrm{r}^\dagger a_\mathrm{q}^\dagger$ and $a_\mathrm{r}a_\mathrm{q}$ have different signs with respect to each other. Here, it seems that the type of coupling could lead to different energy spectra.

Before we delve further into this question, the second approach of diagonalizing the Hamiltonian must be briefly mentioned, which was the first approach presented in the literature [219]. Using an orthogonal transformation $\mathbf{S}$ that diagonalizes $\hbar^2\boldsymbol{\omega}^2$, one obtains

$$H = \frac{1}{2\hbar^2}\mathbf{p}'^T\mathbf{p}' + \frac{\hbar^2}{2}\mathbf{x}'^T \begin{pmatrix} \omega_1'^2 & 0 \\ 0 & \omega_2'^2 \end{pmatrix} \mathbf{x}' - E_\mathrm{J}\cos(ux_1' + vx_2')\,. \tag{6.23}$$

Here, the primed variables denote the coordinates, u and v are scalars resulting from the coordinate transformation, and ω_1' and ω_2' are the frequencies of the resonator and the qubit in the absence of the nonlinear interaction, i.e. with $E_\mathrm{J} = 0$. Depending on the degree of coupling it might no longer be meaningful to identify qubit and resonator directions, which is why the coordinates are simply denoted as x_1' and x_2'. The advantage of this approach is that the linear Hamiltonian is diagonal and the perturbation lies solely in the nonlinear potential. However, a clear disadvantage of this approach is that the Laguerre polynomials for the resonator states have to be evaluated, which is computationally expensive, especially when high photon numbers in the resonator are to be investigated.

6.3 Mapping between inductive and capacitive coupling

In the following, it will be shown that the inductive and capacitive coupling can result in the same system Hamiltonian. As was briefly mentioned in Sec. 5.1.2, the idea is to transition to a coordinate system where the particle has the same effective mass $\hbar^2$ in all dimensions. In principle, this means to introduce the coordinates $\mathbf{x} = 2\pi/\Phi_0 \cdot \mathbf{E}_\mathrm{C}^{-1/2}\boldsymbol{\phi}$ and $\mathbf{p}^T = \mathbf{q}^T\Phi_0/2\pi \cdot \mathbf{E}_\mathrm{C}^{1/2}$. However, as for real numbers, computing the square root of a matrix is not a unique matter, since any $\mathbf{E}_\mathrm{C}^{1/2} \to \mathbf{E}_\mathrm{C}^{1/2}\mathbf{S}$ with $\mathbf{S}$ being orthogonal is equally suited. For the inductive coupling, the natural choice is to compute element-wise the square root of $\mathbf{E}_\mathrm{C}$. The Hamiltonian then becomes

$$H = \frac{1}{2\hbar^2}\mathbf{p}^T\mathbf{p} + \frac{\hbar^2}{2}\mathbf{x}^T\mathbf{E}_\mathrm{C}^{1/2}\mathbf{E}_\mathrm{L}\mathbf{E}_\mathrm{C}^{1/2}\mathbf{x} - E_\mathrm{J}\cos\left(\sqrt{E_{\mathrm{C_q}}}x_\mathrm{q}\right). \tag{6.24}$$

In this new coordinate system, we can now define the omega matrix as

$$\hbar^2\omega^2 := \mathbf{E}_\mathrm{C}^{1/2}\mathbf{E}_\mathrm{L}\mathbf{E}_\mathrm{C}^{1/2} = \hbar^2\begin{pmatrix}\omega_\mathrm{r}^2 & -g^2\\ -g^2 & \omega_\mathrm{q}^2\end{pmatrix}. \tag{6.25}$$

Note that due to the squared energies, g^2 does not correspond to the frequency splitting of the harmonic modes at the avoided level crossing. Instead, when $\omega_\mathrm{r} = \omega_\mathrm{q} := \omega$, the eigenfrequencies are $\omega_{1,2} = \sqrt{\omega^2 \pm g^2} \approx \omega \pm g^2/(2\omega) - \cdots$.

Can the Hamiltonian of the capacitive coupling be mapped onto the Hamiltonian of the inductive coupling? The naive approach would be to first diagonalize the capacitance matrix $\mathbf{E}_\mathrm{C} = \mathbf{SDS}^T$ and to compute the square root via $\mathbf{E}_\mathrm{C}^{1/2} = \mathbf{SD}^{1/2}\mathbf{S}^T$. The resonator coordinate would then appear in the cosine potential and would have to be removed by an additional rotation. The choice of keeping the direction of the fluxonium unchanged allows for a simpler and even algebraic transformation. This strategy is also interesting for even higher dimensional systems, where the diagonalization of $\mathbf{E}_\mathrm{C}$ can no longer be calculated analytically. The idea is to compute the unique Cholesky decomposition of $\mathbf{E}_\mathrm{C} = \mathbf{L}\mathbf{L}^T$ with $\mathbf{L}$ being an upper triangular matrix. One obtains

$$\mathbf{L} = \begin{pmatrix} \sqrt{E_{\mathrm{C_r}} - \frac{E_{\mathrm{C_s}}^2}{E_{\mathrm{C_q}}}} & \frac{E_{\mathrm{C_s}}}{\sqrt{E_{\mathrm{C_q}}}} \\ 0 & \sqrt{E_{\mathrm{C_q}}} \end{pmatrix}. \tag{6.26}$$

Now, we can define $\mathbf{x} = \mathbf{L}^{-1}\boldsymbol{\phi}$ and $\mathbf{p} = \mathbf{L}^T\mathbf{q}$ and obtain

$$\hbar^2\omega^2 = \mathbf{L}^T\mathbf{E}_\mathrm{L}\mathbf{L} = \begin{pmatrix} E_{\mathrm{L_r}}E_{\mathrm{C_r}} - \frac{E_{\mathrm{L_r}}E_{\mathrm{C_s}}^2}{E_{\mathrm{C_q}}} & E_{\mathrm{L_r}}E_{\mathrm{C_s}}\sqrt{\frac{E_{\mathrm{C_r}}}{E_{\mathrm{C_q}}} + \frac{E_{\mathrm{C_s}}^2}{E_{\mathrm{C_q}}^2}} \\ E_{\mathrm{L_r}}E_{\mathrm{C_s}}\sqrt{\frac{E_{\mathrm{C_r}}}{E_{\mathrm{C_q}}} + \frac{E_{\mathrm{C_s}}^2}{E_{\mathrm{C_q}}^2}} & E_{\mathrm{L_q}}E_{\mathrm{C_q}} + \frac{E_{\mathrm{L_r}}E_{\mathrm{C_s}}^2}{E_{\mathrm{C_q}}} \end{pmatrix} \tag{6.27}$$

as well as the nonlinear potential

$$-E_\mathrm{J}\cos\left(\sqrt{E_{\mathrm{C_q}}}x_q\right). \tag{6.28}$$

The sign difference in the off-diagonal entry of the omega matrix can simply be removed by changing the sign of the resonator axis. By choosing the right energies $E_{\mathrm{L_r}}$, $E_{\mathrm{L_q}}$, and $E_{\mathrm{C_s}}$ this matrix can coincide with the omega matrix for the inductive coupling (s. Eq. 6.25). We therefore obtain the same energy spectrum of the resonator and fluxonium. However, the matrix elements for the charge and flux fluctuations are most likely different. This means that one of the coupling approaches may be superior when it comes to optimizing dissipation.

6.4 Sorting of the states

Diagonalization of the coupled resonator fluxonium Hamiltonian gives a complex energy spectrum, as can be seen in Fig. 6.2a. As mentioned in the beginning of this chapter, the states should ideally be sorted for excitations in the resonator and the fluxonium. Clearly, for a strong coupling, the sorting of the states is no longer meaningful due to the overlap of the states. For weak coupling, a first idea would be to compare the states with the product states of the uncoupled resonator fluxonium system, for example by finding the states with the maximal overlap. It turns out that this approach fails already for very weak coupling. In addition, this sorting also lacks the connection to the experiment, since nature can not know about the underlying product basis. It is therefore desirable to think of a sorting that is based on the eigenstates and mimics the experiment.

When the drive of the resonator is turned on, neighboring resonator states will undergo Rabi oscillations. The matrix elements between the states determine the speed of the Rabi oscillations and on which states the resonator climbs up the excitation ladder during the displacement pulse. In steady-state and depending on

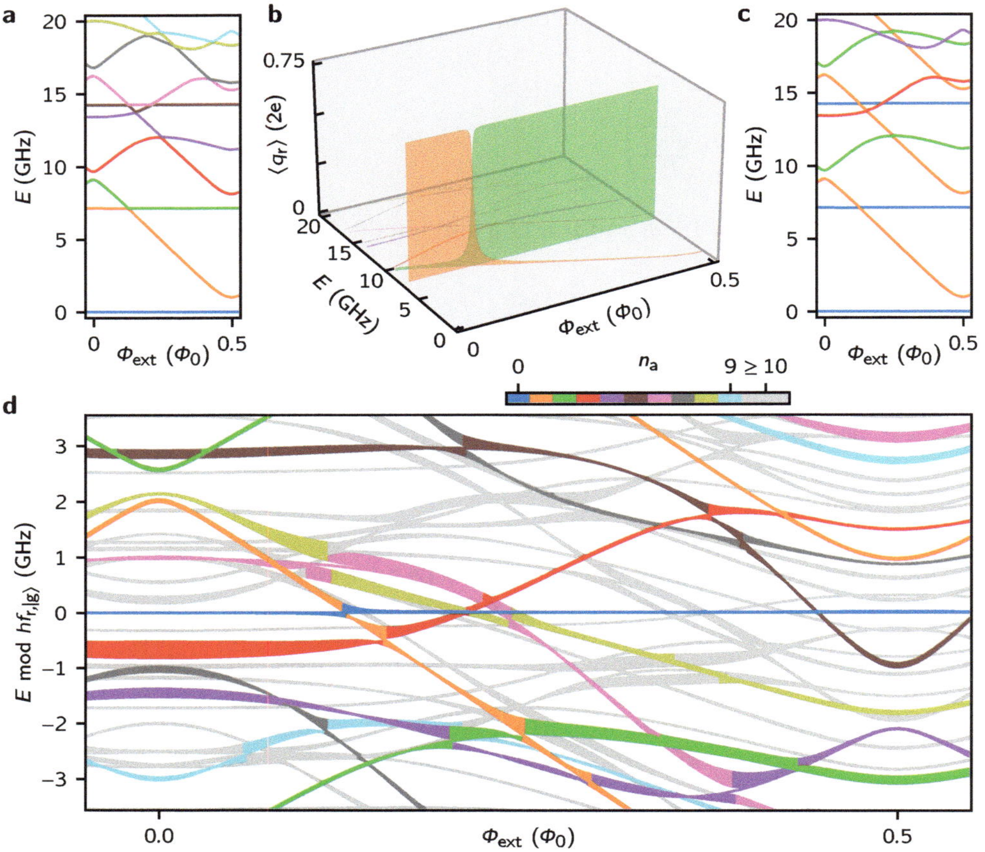

Figure 6.2: Sorting the spectrum. a Unassigned energy spectrum of the resonator fluxonium Hamiltonian as a function of the external flux. **b** Resonator charge dipole moment between the ground state $|0,0\rangle$ and all yet unassigned states, corresponding to the energy spectrum shown in panel a. The first excited resonator state $|1,0\rangle$ can unambiguously be identified. **c** Assigned energy spectrum using different colors for the qubit states. For the parameters see Tab. 6.1. **d** Assigned spectrum modulo the resonator frequency in the ground state, showing the lowest 250 states with energies up to 120 GHz and thus up to 28 qubit and 16 resonator excitations. The figure was created with the bfqcircuits library [237].

the initial qubit state, the resonator will end up in a displaced coherent state, which can be a highly deformed state in phase space [40].

A much more robust sorting approach is therefore the following: we search for the next yet unassigned state $|k,l\rangle$, which has the highest dipole moment with the current state $|m,n\rangle$. Since the resonator is usually driven capacitively, the absolute value of the charge dipole moment $|q_r^{\text{ZPF}} \langle m,n|a_r^\dagger - a|k,l\rangle\,|$ is computed. The dipole moments are depicted in Fig. 6.2b. Starting in the ground state $|0,0\rangle$, we climb up the resonator

Table 6.1: Parameters for studying the coupled resonator fluxonium system. These parameters have been targeted for the fabrication of the granular aluminum fluxoniums.

resonator	fluxonium	coupling
$L_r = 25\,\mathrm{nH}$ $C_r = 20\,\mathrm{fF}$	$L_q = 300\,\mathrm{nH}$ $C_q = 5\,\mathrm{fF}$ $E_J = 9\,\mathrm{GHz}$	$L_s = 3\,\mathrm{nH}$

Table 6.2: Parameters used for generating Fig. 6.3b. The values had to be slightly adjusted compared to those extracted from the spectrum in Fig. 5.15.

resonator	fluxonium	coupling
$L_r = 22.0\,\mathrm{nH}$ $C_r = 22.0\,\mathrm{fF}$	$L_q = 233.7\,\mathrm{nH}$ $C_q = 6.90\,\mathrm{fF}$ $E_{J_i} + E_{J_o} = 20.50\,\mathrm{GHz}$ $E_{J_o} - E_{J_i} = 0.62\,\mathrm{GHz}$ $r = 50.0$	$L_s = 0.64\,\mathrm{nH}$

excitation ladder until the highest computed resonator photon number is reached. Then, we start all over again with the lowest yet unassigned state, which must be $|0, 1\rangle$, the first excited state of the fluxonium qubit. This recipe is continued until all states of the spectrum are assigned. The sorting can be accelerated by searching the next state only within a small energy window corresponding to the next resonator photon energy. In case all states have been assigned in this energy interval, the sorting can no longer be trusted beyond this point. At the end of sorting, an energy limit can be returned, below which the state assignment can be trusted. The sorted states are depicted in Fig. 6.2c. A convenient way to display the spectrum for large energies, is to plot the spectrum modulo the resonator frequency as presented in Fig. 6.2d, showing the spectrum with up to 28 qubit and 16 resonator excitations. For a high fidelity readout avoided level crossings should be avoided. For instance, the eights qubit state at half flux in Fig. 6.2d gets dangerously close to the first excited state.

Exactly the same sorting strategy has recently been used to explain the ionization of transmon qubits during the readout [40], which has puzzled the community almost since the invention of the dispersive readout scheme [34, 35].

The resonance frequency of the resonator is likely to branch at higher photon numbers due the increasing number of avoided level crossings and hybridizations with nearby levels. This effect can also be observed for our fluxonium. In Fig. 6.3a, the measured resonator response of the SQUID fluxonium is shown. Here, the qubit is mostly in

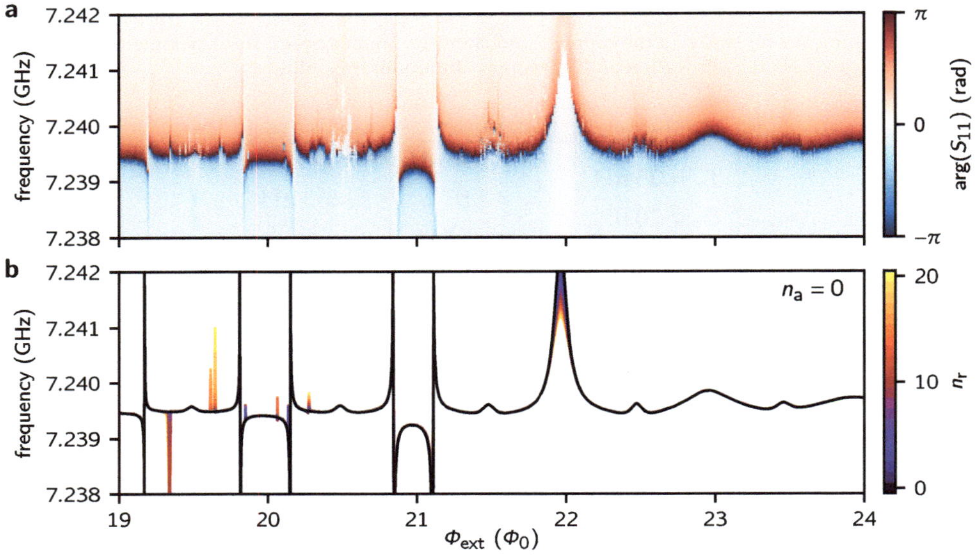

Figure 6.3: Resonator response as a function of the external flux. a Measured resonator response of the SQUID fluxonium as a function of the external flux and the drive frequency. The measured fluxonium spectrum is shown in Fig. 5.15. **b** Calculated resonator frequency $f_{\mathrm{r},|g\rangle}(n_\mathrm{r}) = \omega_{\mathrm{r},|g\rangle}(n_\mathrm{r})/2\pi$ for the qubit being in the ground state, i.e. $n_\mathrm{a} = 0$. For the parameters see Tab. 6.2.

the ground state, except when the external flux is close to half flux integer values, where the qubit is also thermally excited (cf. energy spectrum in Fig. 5.15). In the resonator response fine features are visible, some of which are also predicted by the sorting algorithm. Since the algorithm selects only the state with maximal dipole overlap, not all features are detected. For a quantitative agreement, the full resonator response must be computed quantum mechanically [40].

The spectrum at large photon numbers might explain the often exploited sideband cooling and population inversion with large photon numbers at particular flux spots, as can be seen in Fig. 6.6 in the next section. As was already mentioned above, for a high fidelity qubit readout, it is therefore important to design the fluxonium spectrum in such a way that no avoided level crossings or nearby levels are present in the spectrum up to large photon numbers.

6.5 Dispersive shift

The dispersive frequency shift χ of the resonator induced by the fluxonium inherits the complexity of their joined energy spectrum. Due to the avoided level crossings in

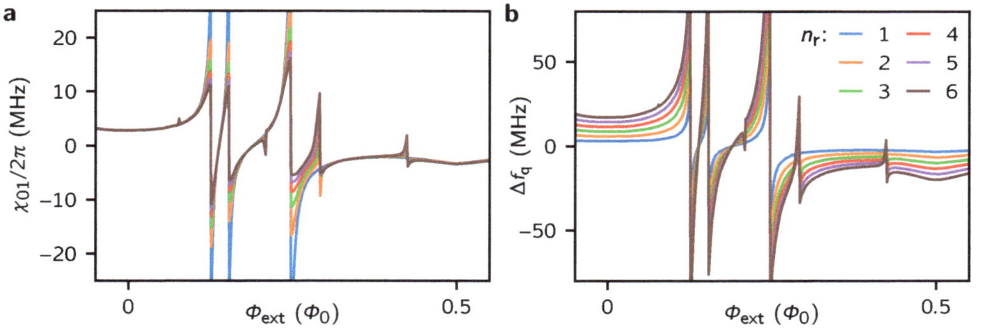

Figure 6.4: Dispersive shift of the fluxonium. **a** Dispersive shift χ_{01} and **b** frequency shift Δf_q of the fluxonium qubit as a function of the external flux and for various photon numbers. For parameters see Tab. 6.1.

the spectrum, the dispersive shift is likely to undergo abrupt changes as a function of the external flux, as can be seen in Fig. 6.4a, where the dispersive shift of the fluxonium qubit is shown. The frequency shift Δf_q of the qubit is shown in Fig. 6.4b for various photon numbers n_r in the resonator. Here, we define the shift relative to the bare qubit frequency when the resonator is in the ground state. Thus,

$$h\Delta f_q(n_r) = (E_{|n_r,1\rangle} - E_{|n_r,0\rangle}) - (E_{|0,1\rangle} - E_{|0,0\rangle}). \tag{6.29}$$

The amount of features in the dispersive shift increases with the number of photons in the resonator, as can be seen in Fig. 6.4. If the parameters are tuned in the right way, these frequency collisions can to a large extent be avoided. In this case, the dispersive shift of the resonator changes rather smoothly with increasing photon numbers. In Fig. 6.5, the dispersive shift of the resonator is depicted for the quantum Rabi and the Jaynes-Cummings model (panel a), and for the fluxonium (panel b) as a function of the number of photons in the resonator. For the fluxonium, we show in addition the dependence for the higher fluxonium states $|f\rangle$ and $|h\rangle$, corresponding to $n_a = 2$ and $n_a = 3$. While the Jaynes-Cummings model predicts a rather slow dependence of the dispersive shift with n_r, the dispersive shifts from the quantum Rabi model and induced by the fluxonium vary much faster and $\chi_{01}/2\pi = f_{r,|e\rangle} - f_{r,|g\rangle}$ undergoes several changes in sign. Interestingly, the pull of the fluxonium ground state has almost no power-dependence for more than 200 photons in the resonator.

This complex dependence is certainly challenging, but also bears great potential, since the spectrum can be engineered to yield an almost perfect QND Hamiltonian (Eq. 6.1) that is valid for large photon numbers and should allow for a very fast readout with high fidelity. For instance, one may replot Fig. 6.5b with $E_J = 11$ GHz, which yields almost no photon number dependence for the ground and excited state. An experimental proof for the existence of high fidelity readouts in fluxonium qubits using large photon numbers has been demonstrated with our granular aluminum

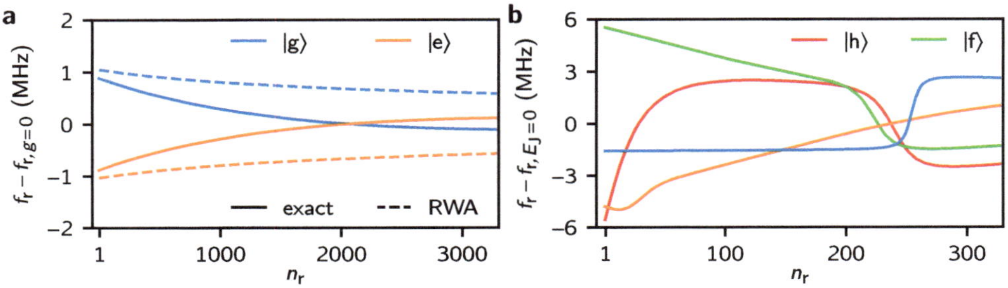

Figure 6.5: Power dependence of the dispersive shift. a Dispersive frequency shift of the resonator for the qubit being in the ground or excited state with respect to the resonator's bare frequency $f_{r,g=0}$ as a function of the photon number n_r in the resonator. The continuous lines depict the exact result of the quantum Rabi model, while the dashed lines correspond to the Jaynes-Cummings model (Eqs. 6.9 and 6.11), which rests on the rotating-wave approximation (RWA). The resonator and qubit frequencies, $f_{r,g=0} = 7.12\,\text{GHz}$ and $f_q = 978\,\text{MHz}$, respectively, correspond to those in panel b. For the quantum Rabi and Jaynes-Cummings model $g = 2\pi \cdot 150\,\text{MHz}$ and $g = 2\pi \cdot 80\,\text{MHz}$ was chosen, respectively, to yield a dispersive shift of $\chi_{01} \approx 2\pi \cdot 2\,\text{MHz}$. **b** Dispersive frequency shift of the resonator in dependence of the resonator photon number for the lowest four fluxonium levels at half flux. For the parameters see Tab. 6.1. The bare frequency $f_{r,E_J=0}$ with vanishing Josephson energy $E_J = 0$ was subtracted from the resonator frequency f_r. The calculations were performed using $N_r = 380$ and $N_a = 50$ harmonic oscillator states for the resonator and the fluxonium.

fluxonium [33]. Note that in reality, there are various spurious modes besides the resonator mode that complicate the situation. The use of granular aluminum instead of junction arrays (s. Sec. 5.2.2) is certainly advantageous here, since the plasma frequency of granular aluminum lies at much higher frequencies than the one of junction arrays, resulting in fewer spurious modes in the relevant frequency range.

The dispersive shift was measured experimentally by recording continuous quantum jump traces at moderate photon numbers using a parametric amplifier. For comparison, the dispersive shift is also calculated with the numerical model. Fig. 6.6 shows the measured and calculated dispersive shift of the SQUID fluxonium in dependence of the external flux close to four different half flux integer values. The measured dispersive shift can be well described by the numerical model for all four half flux integer values.

Interestingly, it can be seen in the measurements that the thermal qubit population does not show a smooth behavior over flux, i.e. with increasing qubit frequency. Instead, distinct features appear, showing effects of side-band cooling and population inversion. Since the measurement was carried out at moderate photon numbers, these features are likely caused by avoided level crossings in the spectrum at higher photon numbers. However, the fact that the amount of features increases with decreasing qubit frequency in the four panels could also indicate the presence of other effects in the environment. Another remaining puzzle is the vanishing dispersive shift of the excited state in the upper left panel, which is not predicted by the theory. However,

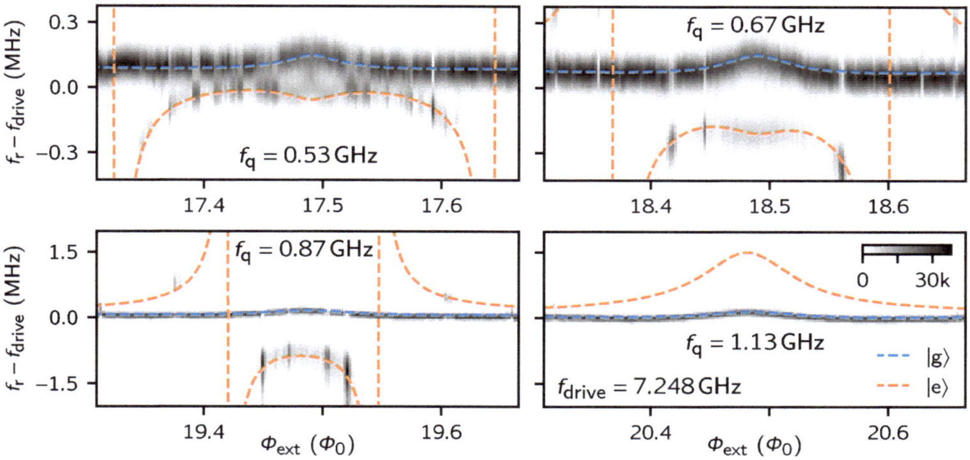

Figure 6.6: Measured dispersive shift close to half flux integer values. Measured (black histogram) and calculated (dashed lines) dispersive shift of the SQUID fluxonium as a function of the external flux Φ_{ext} close to different half flux integer values for each of the panels. The qubit frequency at half flux is given in each panel. For the corresponding fluxonium spectrum and the fluxonium parameters see Fig. 5.15. The dispersive shift was measured by recording continuous quantum jump traces at moderate photon numbers using a parametric amplifier. The measured phase $\arg(S_{11})$ was binned in one degree steps and transformed into frequency. The thermal qubit population is overlayed by features showing effects of side-band cooling and population inversion, which are likely caused by avoided level crossings in the spectrum. Interestingly, in the upper left panel, the measured dispersive shift of the excited state vanishes close to the qubit's half flux, which is not predicted by the theory. A possible explanation could be a vanishing qubit lifetime. In all panels, a drive frequency of f_{drive} = 7.248 GHz was used to probe the resonator.

this feature could also be an artifact of a vanishing qubit lifetime that falls well below the integration time used in the vector network analyzer. In this case, we can no longer distinguish between the ground state and excited state and an average response is measured.

7 Experimental techniques

This chapter details some of the experimental techniques used in the thesis, including design and fabrication, the sample holder and the magnetic shielding, and the microwave periphery for measurement and control. Although all measurements presented in this work were conducted on granular aluminum (grAl) fluxoniums designed for a microwave waveguide sample holder, experiments were also performed on grAl fluxoniums in a 2D architecture featuring a fast flux line, in collaboration with Patrick Paluch. Both designs and their fabrication will be briefly presented.

7.1 Design and fabrication

The parameters targeted for the fabrication of the waveguide samples, which hold approximately also for the samples in the 2D architecture, are stated in Tab. 6.1. In the following, the choice of these parameters is detailed.

In the previous chapter, the dispersive readout scheme has been discussed. In particular, it was mentioned that for a high quality readout with large photon numbers the energy spectrum has to be optimized to avoid frequency collisions leading to unwanted sideband transitions. For this purpose, it is therefore advantageous to choose a rather high readout frequency compared to the qubit frequency in order to yield a lower mode density. A large resonator qubit detuning also leads to a weaker dependence of the dispersive shift on the photon number in the resonator. An additional and unrelated reason for choosing a high readout frequency is to avoid thermal population of the resonator, which would otherwise lead to photon shot noise dephasing of the qubit. For this reason, the readout frequencies are typically chosen above 5 GHz. Our waveguide sample holder (Sec. 7.2) is impedance matched in the 6-8 GHz frequency range. Readout frequencies in this range were also targeted for the 2D architecture, even though this architecture comes without additional frequency restrictions.

The coupling of the readout resonator to the microwave transmission line was engineered to yield a resonator linewidth of $\kappa/2\pi \sim 1\,\mathrm{MHz}$, which allows for a fast ring up and ring down of the resonator photons within a few hundred nanoseconds. While a fast ring up can always be achieved by increasing the input power, a fast ring down with a phase controlled cancellation pulse [45] ideally requires knowledge about the qubit state [46], limiting the total readout pulse length, in this case to at least the latency of the readout electronics. Choosing $\kappa > 2\pi \cdot 1\,\mathrm{MHz}$ would yield an even faster resonator population. However, if the dispersive shift χ_{01} of the qubit is not increased accordingly, the readout signal vanishes. The readout signal is optimized by designing the dispersive shift χ_{01} of the fluxonium qubit biased at half flux such that $\chi_{01} \sim \kappa$. Further increasing χ_{01} does not improve the readout signal but still increases the photon shot noise induced qubit dephasing. Depending on the design, a large dispersive shift can also degrade the qubit lifetime due to the Purcell loss, i.e. radiation loss of the qubit via the resonator into the microwave lines. For the microwave waveguide, which has a cutoff frequency of 6 GHz, the Purcell loss into the microwave lines can be neglected. If, however, the resonator has intrinsic losses due to other broadband environments, the qubit will inherit these losses in dependence of the qubit resonator coupling.

We decided for the inductive resonator fluxonium coupling scheme. The main advantages is that no additional capacitor pads at the Josephson junction are needed, which are prone to dielectric losses from to surface TLSs [65]. This argument is one of the motivations for the development of mergemons [238, 239]. Overall, the fluxonium and resonator parameters that we converged on are similar to designs reported in the literature [71, 240].

The first design idea was to produce the fluxonium, the superinductor and the junction entirely out of granular aluminum. The antenna pads for the resonator would have been added in a second optical lithography step. Since it was rather difficult to get the right junction resistance, this design was given up. Instead, a three-angle shadow evaporation process was developed to implement both a conventional $Al/AlO_x/Al$ Josephson junction and the grAl superinductance in a single electron beam lithography step.

The main steps of the fabrication process are illustrated in Fig. 7.1. A table with the detailed fabrications steps is given in App. G. The three-angle process employs a commonly used bi-layer resist stack [241] on top of a c-plane sapphire wafer. For the bottom layer, the EL-13 copolymer MMA resist is used (MicroChem) and spin-coated to a thickness of 950 nm. The second layer consists of the A4 PMMA resist (MicroChem), here a thickness of 200 nm is targeted. Finally, a thin gold layer is sputtered on top for the subsequent e-beam writing (Fig. 7.1a). The e-beam lithography system (JEOL) that writes the desired pattern into the resist uses an acceleration voltage of 50 kV. Due to the rather low acceleration voltage, the electrons scatter in the resist, which leads to a broadening of the exposed area towards the substrate. In addition, the electrons reflect from the substrate and create a second undercut after the development of the resist. After the e-beam exposure the gold layer is removed with a 15 % Lugol solution and the resist is developed in an isopropanol (IPA) water solution with ratio 3:1 cooled to 6 °C to increase the contrast of the two resist layers. This resist stack enables the fabrication of undercuts (Fig. 7.1b) and bridges [241, 242]. In conjunction with a tilted shadow electron beam evaporation of the aluminum and granular aluminum films, this resist stack allows for flexible design patterning. For the lift-off process, the sample is exposed to N-Ethyl-2-

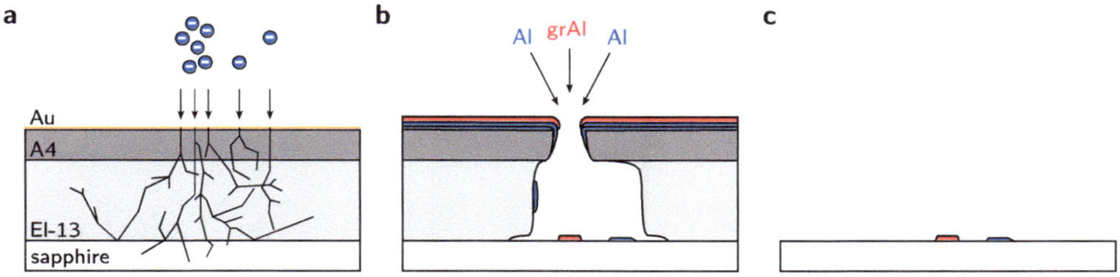

Figure 7.1: Illustration of the fluxonium fabrication by a three-angle shadow evaporation process. **a** E-beam writing of the pattern into the bi-layer resist. The lower resist is more sensitive to the e-beam exposure than the upper resist. This enables the fabrication of bridges (not shown) and undercuts (panel b). **b** Development of the resist and shadow evaporation. The first two aluminum layers are evaporated under ±30° tilt angle. The first layer is oxidized to form the junction barrier. The grAl film is evaporated with zero tilt angle. **c** Lift-off process to remove the resist and all the metal layers attached to it.

pyrrolidone (NEP) at a temperature of 90 °C for a duration of 90 min, followed by rinsing in acetone and IPA.

The single lithography step comes along with a few restrictions on the resist patterning. The superinductor that is implemented with a granular aluminum film must be interrupted by the junction. This essentially demands for a Dolan bridge [242]. Since granular aluminum is prone to quasiparticles [63] and has a higher critical temperature than aluminum, the aluminum junction area could act as a quasiparticle trap, thereby degrading the lifetime of the qubit. The idea was therefore to connect the granular aluminum film to the junction leads far away from the junction. In this way, the quasiparticles can potentially recombine and thermalize before reaching the junction. Note that fluxoniums biased at half flux are only insensitive to quasiparticle losses at the junction if the quasiparticles are thermalized [55]. For the SQUID fluxonium, the destructive quasiparticle interference can not fully be reached, as discussed in Sec. 5.3.2. It would certainly be interesting to implement the junction with materials that have a higher superconducting gap than granular aluminum, e.g. using a Nb/AlO_x/Nb trilayer. This, however, becomes only relevant if the TLS environment has been identified and its dominant influence has been averted.

The junction resist structures that achieve the desired pattern are a combination of a Dolan bridge and undercuts, as shown in Fig. 7.2. The fabrication is now as follows: First, the junction is patterned with two aluminum evaporations from left and right with a tilt angle of ±30°, interrupted by a static oxidation of the first layer. For the oxidation of the SQUID fluxonium, we used 15 mbar for a duration of 5:30 min. In other fabrication runs the duration was adjusted depending on the targeted Josephson energy and junction design. The first layer has a thickness of 20 nm, while the second layer thickness is 30 nm to overcome the height difference. The two aluminum layers also pattern the resonator antenna pads or the capacitor shoe in the

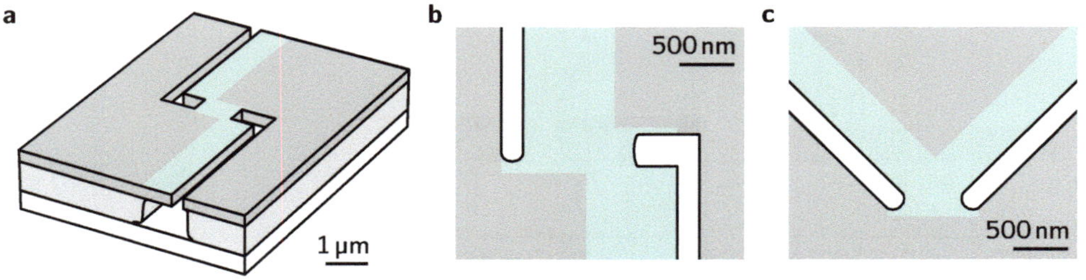

Figure 7.2: Three-angle shadow evaporation. **a** Schematic of the sapphire wafer and resist stack for patterning the Josephson junction of the fluxonium together with its aluminum feedlines. **b, c** Top view of the resist stack as shown in panel a for other pattern geometries, where the junction size is insensitive to height fluctuations of the resist stack. The green area indicates the area where the lower resist has been dissolved. The aluminum evaporation to form the junction is performed from left and right under ±30° tilt angle. The grAl wire byproducts, evaporated with zero tilt angle, do not touch the junction and its feedlines. Reproduced and adapted from Ref. [18].

2D architecture. Next, the grAl film is deposited straight from the top by evaporating aluminum in an oxygen atmosphere. For the sample on which the hyperpolarization experiment was carried out, a thickness of 40 nm was deposited at a rate of 0.3 nm/s to overcome the thickness of the previous Al layer. The resistance of the grAl film fluctuates strongly between different depositions, since it is exponentially sensitive to the dynamical oxygen pressure [17], which builds up by the combination of oxygen supply, pumping of the chamber, and the oxidation of the evaporating aluminum. In subsequent developments, the evaporation rate was increased to 1 nm/s, thereby greatly reducing the resistivity fluctuations. In addition, an in situ resistance measurement device was integrated, which allows adapting the thickness and thus the sheet resistance to further compensate fluctuations. A proper connection to the aluminum wires is ensured by additional contact pads that will be discussed below.

The junction pattern depicted in Fig. 7.2a shows the resist structure that was used to fabricate the sample on which the hyperpolarization experiment was carried out. The final junction patterns can be seen in Fig. 7.3 and Fig. 7.4. In later fabrications, the structure was optimized as illustrated in Fig. 7.2b, c to yield a junction area that is independent of the exact height of the resist stack.

A potential disadvantage of the presented designs is that the junction is hidden under the bridge. The cleaning of the wafer from residual resist by a plasma descum process prior to the evaporation of the metals may not clean the junction area that effectively. However, we did not observe any significant junction aging over the years. A second disadvantage of the three angle approach is that the granular aluminum wires can not be freely patterned. The width of the grAl wires is limited to ~ 300 nm. The shared inductance in Fig. 7.2b has a width of 250 nm. The maximum width can in principle be increased with a higher resist stack, larger

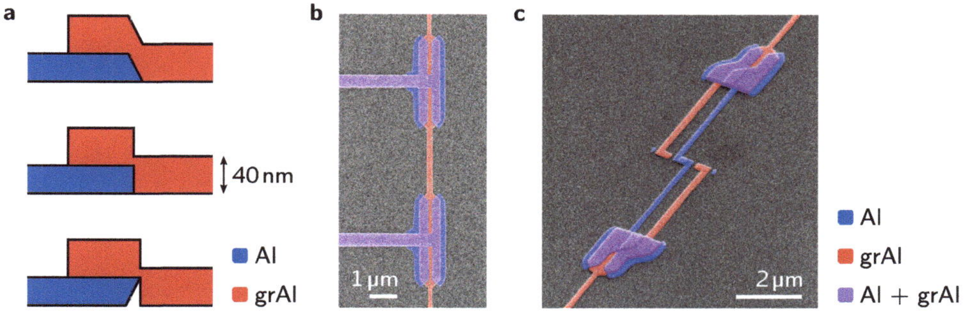

Figure 7.3: Contact pads for secure contacting of grAl and aluminum layers. **a** Idealized illustration of the shadow evaporation of the 30 nm thick aluminum layer and the 40 nm thick grAl layer evaporated straight from the top. The aluminum evaporation can either appear from the left (top), perpendicularly to the drawing plane (middle), or from the right (bottom). Only in the top scenario a continuous growth of the grAl film is ensured. The other scenarios are likely to give constrictions. **b** Contact pads of the shared inductance implementing the top scenario of panel a. The shared inductance has a width of 250 nm and the inductance of the superinductor has a width of 170 nm. **c** Contact pads for the junction realizing the top scenario of panel a. If no argon milling of the contact pads is applied before the grAl evaporation, the area of the parasitic junction should be enlarged by increasing the length of the contact pads. Regions where the grAl film is shunted by aluminum underneath are false-colored in violet.

angles, and with a 100 keV e-beam writer, where the scattering of the electrons in the resist is less pronounced. Moreover, wires in horizontal direction are shunted by the aluminum layers underneath (Fig. 7.3 and Fig. 7.4). Nevertheless, a zick-zack meandering of the superinductor under 45° angles would in principle be possible. This would allow reducing the area of the superinductor loop. In this way, the frequencies of the spurious modes of the superinductor loop can be increased. We decided for a simple design process, considering the fact that the spurious modes of the superinductor loop have been simulated to be above 14 GHz for the targeted inductance of $L_q = 300\,\mathrm{nH}$.

In order to ensure a good connection from the grAl film to the aluminum layers, connection pads have been added. The main purpose of the pads is not only to reduce a potential contact resistance but also to ensure a continuous grAl film over the step of the aluminum layers (s. Fig. 7.3a). The schematics of the layers is certainly idealized. In reality, the edges are presumably rounded. Nevertheless, ideally the top scenario of Fig. 7.3a is realized, where the grAl film grows over the averted edge of the aluminum film. This should be implemented for both aluminum layers. The contact pads in Fig. 7.3b, c have been designed to realize this scenario. In later developments, an argon milling step was introduced before the grAl evaporation to remove the oxide of the lower aluminum layer resulting from the junction oxidation step at the contact pads. In addition, the oxygen flow for the grAl evaporation was added to the aluminum evaporation shortly ($\sim 10\,\mathrm{s}$) before opening the shutter in

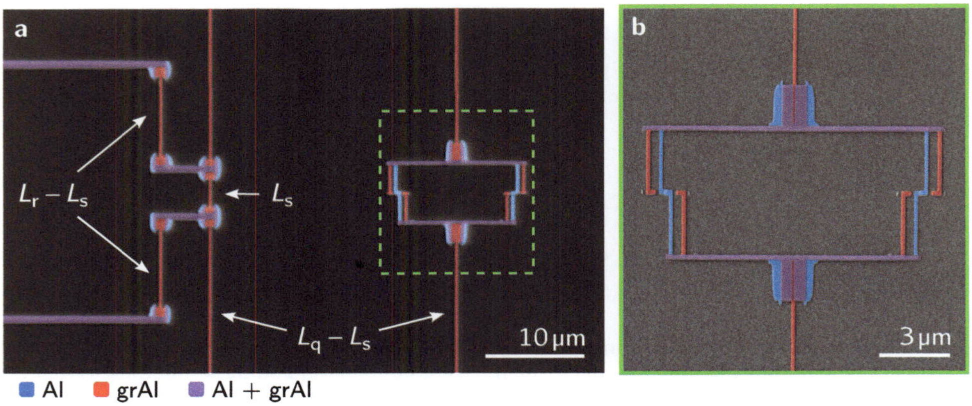

Figure 7.4: Granular aluminum fluxonium design. a Close-up of the false-colored optical image of the granular aluminum fluxonium shown in Fig. 2.6. Indicated are the resonator inductance L_r, the shared inductance L_s, and the qubit inductance L_q. **b** False-colored SEM image of the SQUID. Regions where the grAl film is shunted by aluminum underneath are false-colored in violet.

front of the sample. Simultaneously, the rate control was adapted to the new mass flow such that the rate is stable again when the shutter opens.

A close-up of the false-colored optical image of the SQUID fluxonium depicted in Fig. 2.6 is shown in Fig. 7.4, together with a false-colored scanning electron microscope (SEM) image of the SQUID.

7.2 Sample holder

The first grAl fluxoniums were fabricated for the waveguide sample holder. Here, the samples can be fabricated in a single lithography step. The interior of the waveguide sample holder can be seen in Fig. 7.5a, showing a cut through the 3D model of the high frequency simulation software (HFSS) from Ansys. The microwave readout port to the left is impedance matched to the waveguide, which has a cutoff frequency at around 6 GHz. The brass screws are used to optimize the transmission in the 6-8 GHz range. In practice, two identical waveguide sample holders are clamped together without their copper lids and the microwave transmission is optimized with the screws. The screws are then fixed with silver paste to stay in position. The quality factor of the readout resonator is simulated with HFSS to yield the corresponding $\kappa \sim 1$ MHz linewidth. In order to drive the qubit at frequencies below the waveguide cut-off, a drive port is needed that couples weakly to the readout antenna. The wafer with the samples is diced to 10 mm × 15 mm chips. On the chip, we typically place four fluxonium resonator circuits 2 mm apart with readout frequencies separated

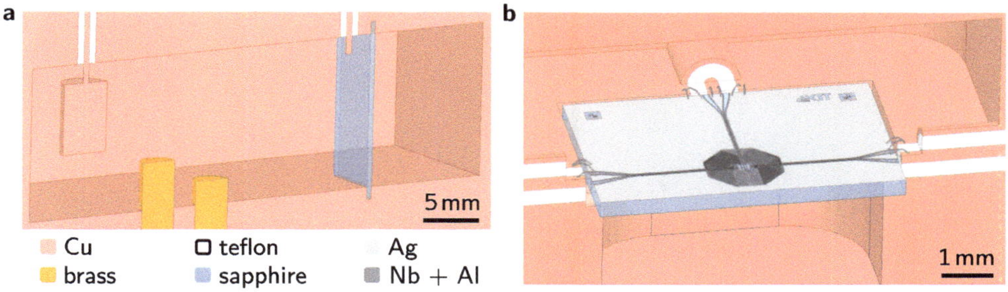

Figure 7.5: HFSS models of the sample holders. a Waveguide sample holder with impedance matched readout port to the left and qubit drive port to the right. The brass screws are used for impedance matching. The chip sits a quarter wave length away (7 GHz) from the end of the waveguide at the antinode of the electric field. **b** Sample holder for two fluxoniums in a 2D architecture. The ports to the left an right are for the fluxonium readout and the port in the middle is used for fast-flux biasing and qubit driving. Only the inner part of the ground plane is superconducting, the remaining part is a thick normal conducting silver ground plane. Not shown are the screws fastening the wafer in the corner and all the wire bonds to the cooper box ensuring a proper ground.

by ~ 150 MHz by simply changing the grAl wire length of the resonator inductor. The chip sits a quarter wave length (7 GHz) away from the end of the waveguide at the antinode of the electric field. It is thermally anchored and held in place with silver paste or vacuum grease. Two of the four SQUID fluxoniums came out with symmetric SQUIDs, which made it possible to tune the Josephson energy to lower values and to enter an suitable fluxonium regime. The hyperpolarization of the TLSs was mainly investigated on these fluxonium samples (s. Supp. Ref. [21]).

The 2D architecture for superconducting qubits requires a lot more engineering than the 3D waveguide. However, the benefit is a significant gain in experimental freedom. In particular, the ability of designing local fast-flux lines enables many possibilities such as Floquet-engineering of the spectrum [243], fast-flux qubit reset and control [50, 156], as well as the hyperpolarization experiments discussed in the outlook. Moreover, fluxoniums qubits, with their compact footprint, small stray capacitance, large anharmonicity, and state-of-the-art coherence times, render them ideal for multi-qubit experiments within a 2D architecture [244].

The main goal of the project was to gain experience with the 2D architecture and to try a few uncommon concepts. The sample holder and the chip layout for the 2D architecture are shown in Fig. 7.5b. In order to compare the fluxonium performance to the performance in the waveguide, the fluxoniums were implemented in exactly the same way, except that the readout resonator was adapted to the coplanar waveguide (CPW) geometry (Fig. 7.6a and Fig. 7.7a, b). The first challenge is to enable a precise global flux biasing of the fluxoniums with their rather large loop area of 150 µm × 25 µm. For this purpose, it must be ensured that there is no flux trapping and pinning in the vicinity of the fluxonium. A superconducting backside

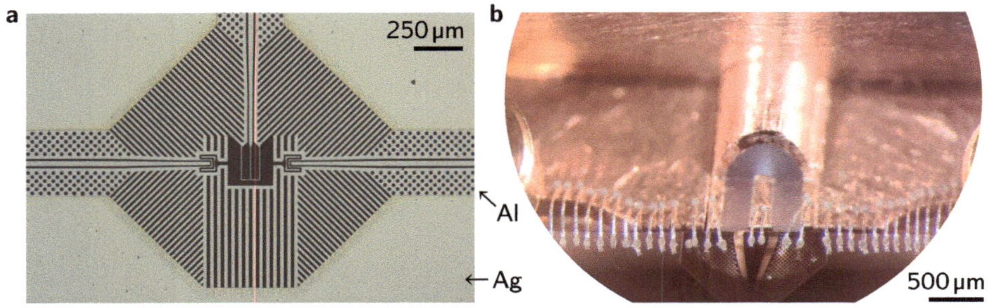

Figure 7.6: Coplanar waveguide design and sample box. **a** Optical image of the first design, distinguished by the flux holes in the CPW. The silver ground plane of this four-year-old sample shows signs of oxidation (yellowish stain). **b** View through the wire bonder showing the microwave taper to the hand-cut coaxial cable.

metalization must therefore be omitted. Similarly, the nearby superconducting ground plane should be a fractal that does not contain superconducting loops or compact superconducting areas larger than the fluxonium area. In order to avoid low frequency modes in the ground plane the fractal needs to be terminated by a normal metal ground plane (Fig. 7.6a and Fig. 7.7a). In addition, the chip is suspended in air to prevent low frequency modes in the substrate (Fig. 7.5b). Furthermore, no printed circuit board (PCB) is used to miniaturize the size of the copper box. Instead, we directly wire-bond the CPWs to the coaxial cables (Fig. 7.5b and Fig. 7.6b). The lowest self resonant modes in the box are the $\lambda/2$ modes of the CPW grounds. If needed, these modes can be shifted to higher frequencies by wire-bonding from ground to ground over the CPW, ideally to the normal metal ground to prevent flux trapping. The termination with the normal metal ground plane has yet other advantages. All the spurious modes in the design that exist at higher frequencies have a low quality factor and are not relevant from a quantum mechanical perspective. In addition, the thick normal metal ground plane acts as a phonon trap, which could reduce the number of excess quasiparticles in the fluxonium [245, 246] and prevent correlated qubit errors [247]. Last but not least, the normal metal ground plane should thermalize the wafer more efficiently, especially when normal metal gold wire bonds are used in addition.

The first implementation can be seen in Fig. 7.6. The design consists of two fluxoniums with corresponding readout resonators, which can be addressed with the left- and right-going CPWs. The fast-flux line in the middle was designed to yield a mutual inductance of 2 pH with the fluxonium loops, such that 1 mA is needed to flux bias the fluxoniums by one flux quantum. The fluxoniums and the resonator inductances are fabricated prior to the ground planes with the three-angle process that was described in the previous section. In the first design, aluminum was used for the superconducting ground plane and the CPW was implemented with flux holes. Since we still observed jumps in the flux biasing of the fluxoniums, the design was

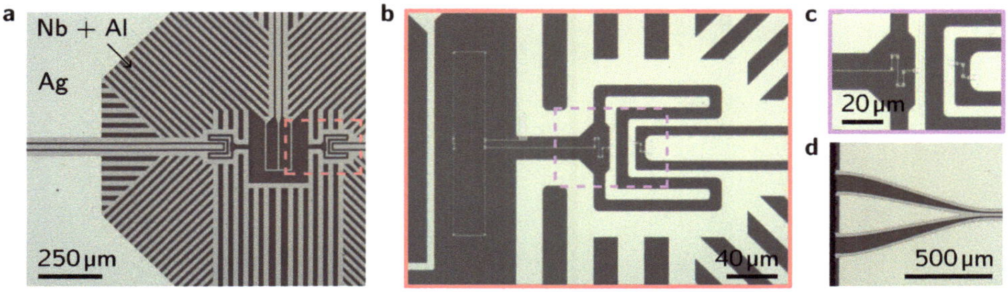

Figure 7.7: Close-up of the coplanar waveguide design. **a** Optical images of the second design, distinguished by the thin superconducting CPW ground and the dark gray niobium aluminum ground plane. **b** Close-up of the fluxonium and its readout resonator. **c** Close-up of the resonator inductance and the $\sim 3\,\text{nH}$ inductive Purcell filter. **d** Microwave taper to the coaxial cable.

adapted for the second implementation, see Fig. 7.7. Here, the flux holes in the CPW ground plane have been removed. It was made sure by simulations that the losses in the CPW are still negligible. In addition, the superconducting ground plane was implemented with 45 nm niobium, evaporated in three 15 nm steps, and 5 nm aluminum on top. The main motivation for adding the aluminum layer on top was to use our calibrated argon milling process for contacting the 200 nm thick silver ground plane in the second optical lithography step. For the silver ground plane, a 4 nm niobium sticking layer was used. In the new design, no flux jumps could be observed anymore.

In the 2D architecture, we measured the longest qubit relaxation times of up to 100 µs. Although these values are due to the frequency characteristics of granular aluminum, as will be shown in a publication in the near future, it shows that the 2D architecture is in no way inferior to the waveguide sample holder. The compact high impedance resonator, along with what is probably the smallest Purcell filter in the literature, may be the future domain of application of granular aluminum. In the presented design, the dispersive shift was difficult to predict, probably due to the asymmetric capacities of the fluxonium loop to the ground plane. In this respect, the design should certainly be improved in the future. Here, the design was intended to be as identical as possible to the design for the waveguide. A potential improvement would be to open the central area for the global flux at the bottom such that the superconducting ground plane is simply connected. This could reduce potential creeping currents in the ground plane.

Magnetic and thermal radiation shielding

The samples must be shielded from magnetic fields and residual thermal radiation from higher temperature stages. First, the magnetic field is attenuated by a μ-

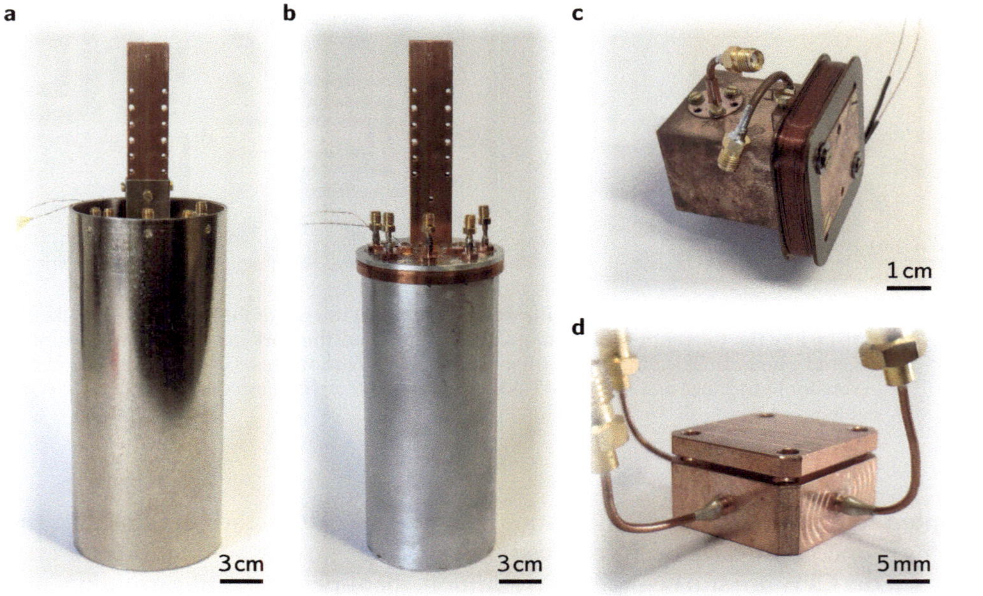

Figure 7.8: Shielding of the samples. **a** Magnetic μ-metal shield. **b** Aluminum shield forced on a copper barrel for magnetic field and thermal radiation shielding **c** Waveguide sample holder with flux bias coil. The microwave port to the left is used for the qubit readout and the port to the right is used for driving the qubit. For further details see Fig. 7.5a. **d** Sample box for the 2D architecture. The port in the middle is used for fast-flux biasing and driving of the two fluxoniums, while the other ports are used for their readout. The microwave cables are fixed with silver paste. For further details see Fig. 7.5b. The global flux bias coil is mounted on top of the copper lid (not shown).

metal shield (Fig. 7.8a). The remaining magnetic field is further shielded and pinned by the superconducting aluminum shield that is forced on a copper barrel (Fig. 7.8b). Only tiny holes remain for the direct current (DC) flux bias lines. This layer therefore provides a good shielding from residual thermal radiation. The waveguide sample holder and the sample box for the 2D architecture are shown in Fig. 7.8c, d, respectively. The sample holder lids are additionally closed with an indium sealing preventing thermal radiation from entering the sample box. The indium sealing was omitted in some of the experiments, which did not seem to harm our samples. However, this may only be true for the rather short ($\sim 10\,\mu s$) coherence time of our samples. Nevertheless, probably the first thing to worry about is the shielding of the microwave cables. The copper post is mounted to the mixing chamber plate of a commercial wet 3He/4He dilution refrigerator (Oxford Instruments Kelvinox 100) reaching $\sim 25\,mK$.

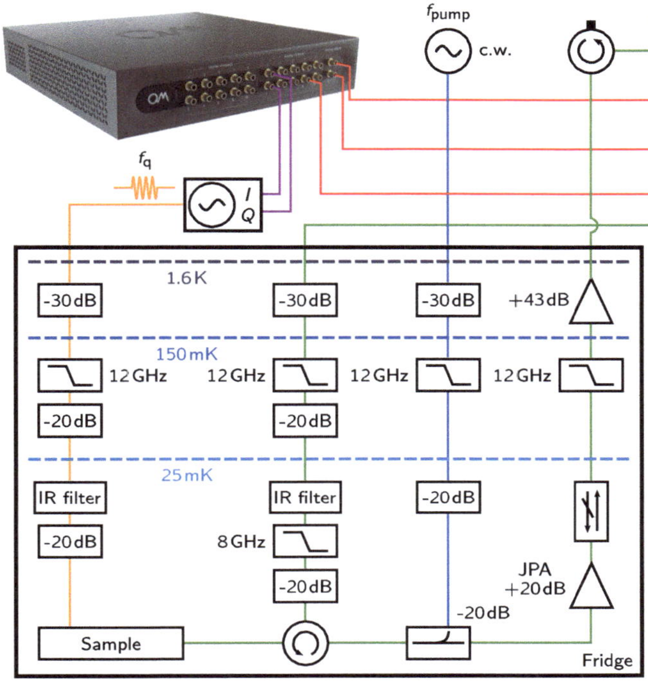

Figure 7.9: Experimental setup for qubit state measurements and active feedback. The OPX instrument from Quantum Machines (top left image) provides the active feedback capabilities. The signal and the reference are interfered computationally by the OPX to extract the I and Q quadratures. We keep the intermediate frequency IF_r of the readout fixed at 62.5 MHz yielding 16 samples per period for the integration. All microwave lines going into the cryostat are attenuated and filtered. The stainless steel microwave line of the readout input (green) provides additional −30 dB attenuation to the 1.6 K-stage (not shown). For more details see text. Reproduced from Ref. [21].

7.3 Microwave setup

In Fig. 7.9, a schematic of the microwave electronics setup for the fluxonium measurement and manipulation is shown. The experimental workflow is orchestrated by the OPX instrument from Quantum Machines (top left image in Fig. 7.9). This FPGA-based instrument can be programmed to measure and estimate the qubit state in real-time using the IQ-demodulated readout signal. Depending on the measurement outcome, it can trigger a π-pulse in order to prepare the qubit in its target state.

The microwave cabling is somewhat more complicated due to an interferometric setup that was required to purify the two-channel microwave generators, which operate in continuous wave (c.w.) mode. The channels use different frequencies to account for the mixers operating on opposite sidebands. Using mixers for the same sideband would work equally well. In later experiments, these sources have been replaced by a single high-quality source with a frequency stability for which phase drifts of the complex scatter parameter are negligible. The attenuation and filtering of the microwave lines going into the fridge is required to shield the sample from thermal radiation. The home-made infrared (IR) filter ensures an attenuation of more than −10 dB for frequencies larger than 60 GHz. This filter was later placed between the sample and the circulator in order to filter also thermal radiation running backwards. The readout signal reflecting from the sample is first amplified with a home-made Josephson junction parametric amplifier (JPA), which provides +20 dB of gain [44]. The pump tone for the JPA is fed into the readout line with a directional coupler providing minimal loss for the readout signal. After the JPA the readout signal passes a 40 dB isolation before it is further amplified by a high-electron-mobility transistor (HEMT). At room temperature, the readout signal is routed through a home-made tunable filter in order to suppress the JPA pump tone, after which it is further amplified, down-converted to the intermediate frequency, and finally recorded.

Appendix

A Notation of the qubit TLS basis functions

The wave function $|\psi\rangle$ of the qubit and the n TLSs can be expressed as $|\psi\rangle = \sum_i c_i |v_i\rangle$, with the basis functions

$$\begin{array}{rclrcl} |v_0\rangle & = & |1111\ldots1\rangle\,, & |v_1\rangle & = & |0111\ldots1\rangle\,, \\ |v_2\rangle & = & |1011\ldots1\rangle\,, & |v_3\rangle & = & |0011\ldots1\rangle\,, \\ \vdots & & & \vdots & & \\ |v_{2^{n+1}-2}\rangle & = & |1000\ldots0\rangle\,, & |v_{2^{n+1}-1}\rangle & = & |0000\ldots0\rangle\,. \end{array} \tag{A.1}$$

Here, the first entry stands for the qubit, the second for the TLS $k = 1$, the third for the TLS $k = 2$, and so forth, with 0 and 1 denoting the ground and excited state, respectively. In this basis, the density matrix reads

$$\rho = |\psi\rangle\langle\psi| = \begin{pmatrix} c_0c_0^* & c_0c_1^* & \cdots \\ c_1c_0^* & c_1c_1^* & \\ \vdots & & \ddots \end{pmatrix}. \tag{A.2}$$

B Liouvillian in superoperator notation

In order to rewrite the Liouville-von Neumann equation in form of a matrix differential equation, the action on each of the density matrix entries must be computed. Following Ref. [137], the trick is to rewrite the equation first in tensor notation. Defining $\vec{\rho} = \sum \rho_{mn} |m\rangle \otimes |n\rangle$, it follows

$$\begin{aligned} (\mathbf{A} \otimes \mathbb{1})\,\vec{\rho} &= \sum_{ij} \rho_{ij} \left(\sum_{nm} \mathbf{A}_{nm} |n\rangle \langle m|i\rangle \right) \otimes |j\rangle \\ &= \sum_{njm} \mathbf{A}_{nm}\rho_{mj} |n\rangle \otimes |j\rangle = \sum_{nj} (\mathbf{A}\rho)_{nj} |n\rangle \otimes |j\rangle = (\vec{\mathbf{A}\rho}) \end{aligned} \tag{B.1}$$

and analogously

$$
\begin{aligned}
(\mathbb{1} \otimes \mathbf{B})\,\vec{\rho} &= \sum_{ij} \rho_{ij} \left|i\right\rangle \otimes \left(\sum_{nm} \mathbf{B}_{nm} \left|n\right\rangle \left\langle m|j\right\rangle \right) \\
&= \sum_{inm} \rho_{im} \mathbf{B}_{nm} \left|i\right\rangle \otimes \left|n\right\rangle = \sum_{in} \left(\rho \mathbf{B}^T\right)_{in} \left|i\right\rangle \otimes \left|n\right\rangle = \left(\overrightarrow{\rho \mathbf{B}^T}\right).
\end{aligned} \tag{B.2}
$$

Similarly, the replacement $(\mathbf{A} \otimes \mathbf{B}^T)\vec{\rho} = (\mathbf{A} \otimes \mathbb{1})(\mathbb{1} \otimes \mathbf{B}^T)\vec{\rho} = \left(\overrightarrow{\mathbf{A}\rho\mathbf{B}^T}\right)$ holds.

Now, using these substitutions and the fact that the Hamiltonian $H = H^T$ is symmetric and the dissipators $L_\alpha = L_\alpha^*$ are real valued, the Lindblad equation (Eq. 3.51) becomes

$$
\overleftrightarrow{\mathcal{L}} = -\frac{i}{\hbar}(H \otimes \mathbb{1} - \mathbb{1} \otimes H) + \sum_\alpha L_\alpha \otimes L_\alpha - \frac{1}{2}(L_\alpha^T L_\alpha \otimes \mathbb{1} + \mathbb{1} \otimes L_\alpha^T L_\alpha). \tag{B.3}
$$

Casting the tensor products in $\overleftrightarrow{\mathcal{L}}$ and $\vec{\rho}$ in matrix and vector form finally gives the desired matrix differential equation, which can then be sorted for diagonal entries and the various coherences.

With the Liouvillian $\overleftrightarrow{\mathcal{L}}$ in tensor form, the matrix elements can easily be computed. For example, for the qubit coupled to two TLSs, one of the off-diagonal entries of $\mathbf{C}_Z$ originates from

$$
\mathcal{L}_{mn} = \left\langle 100\right| \otimes \left\langle 010\right| \overleftrightarrow{\mathcal{L}} \left|101\right\rangle \otimes \left|011\right\rangle = \Gamma_\downarrow^{t_2}. \tag{B.4}
$$

C Dephasing from phase jumps

The statistical density matrix is obtained by averaging over all quantum jump trajectories with their statistical weight. The averaging over the phase jumps can directly be performed so that only the averaging over different numbers n of jumps and their timing remains. Therefore,

$$
\begin{aligned}
\rho(t) = \sum_{n=0}^{\infty} \int_0^t \mathrm{d}t_n \int_0^{t_n} \mathrm{d}t_{n-1} \cdots \int_0^{t_2} \mathrm{d}t_1 e^{-\lambda(t-t_n)} \lambda e^{-\lambda(t_n-t_{n-1})} \ldots \lambda e^{-\lambda(t_1-t_0)} \\
\cdot e^{\mathcal{L}_0(t-t_n)} \langle \Phi_\varphi \rangle e^{\mathcal{L}_0(t_n-t_{n-1})} \cdot \ldots \cdot \langle \Phi_\varphi \rangle e^{\mathcal{L}_0(t_1-t_0)} \rho(t_0).
\end{aligned} \tag{C.1}
$$

The probability $e^{\lambda(t-t_n)}$ ensures that after the nth jump no other jump occurs. Rearranging Eq. C.1 yields

$$
\begin{aligned}
\rho(t) = e^{(\mathcal{L}_0-\lambda\mathbb{1})t} \Big[\sum_{n=0}^{\infty} \int_0^t \mathrm{d}t_n \int_0^{t_n} \mathrm{d}t_{n-1} \cdots \int_0^{t_2} \mathrm{d}t_1 \\
e^{-(\mathcal{L}_0-\lambda\mathbb{1})t_n} \lambda \langle \Phi_\varphi \rangle e^{(\mathcal{L}_0-\lambda\mathbb{1})t_n} \cdot \ldots \cdot e^{-(\mathcal{L}_0-\lambda\mathbb{1})t_1} \lambda \langle \Phi_\varphi \rangle e^{(\mathcal{L}_0-\lambda\mathbb{1})t_1} \Big] e^{-(\mathcal{L}_0-\lambda\mathbb{1})t_0} \rho(t_0).
\end{aligned} \tag{C.2}
$$

The term in the brackets can be recognized as the Dyson series of $\lambda\langle\Phi_\varphi\rangle$ in the interaction picture with respect to the Liouvillian $\mathcal{L}_0 - \lambda\mathbb{1}$. Indeed, taking the derivative shows that $\rho(t)$ solves the differential equation

$$\dot{\rho}(t) = \left(\mathcal{L}_0 - \lambda(\mathbb{1} - \langle\Phi_\varphi\rangle)\right)\rho(t). \tag{C.3}$$

D Approximating the inverse of the coherence matrix

For the derivation of the Solomon equations, we need to compute $\mathbf{C}^{-1}$, where we can make use of the block-wise inversion formula. We write $\mathbf{C}$ as

$$\mathbf{C} = \begin{pmatrix} \mathbf{C}_{\mathrm{ZD}} & \mathbf{S}_{\mathrm{ZD}}^T \\ \mathbf{S}_{\mathrm{ZD}} & \mathbf{C}_{\mathrm{R}} \end{pmatrix}, \tag{D.1}$$

with the index ZD denoting the joint matrices. Under the assumption that the inverse matrices $\mathbf{C}_{\mathrm{ZD}}^{-1}$ and $(\mathbf{C}_{\mathrm{R}} - \mathbf{S}_{\mathrm{ZD}}\mathbf{C}_{\mathrm{ZD}}^{-1}\mathbf{S}_{\mathrm{ZD}}^T)^{-1}$ exist, the upper left block of the inverse matrix $\mathbf{C}^{-1}$ is

$$\mathbf{C}_{\mathrm{ZD}}^{-1} + \mathbf{C}_{\mathrm{ZD}}^{-1}\mathbf{S}_{\mathrm{ZD}}^T\left(\mathbf{C}_{\mathrm{R}} - \mathbf{S}_{\mathrm{ZD}}\mathbf{C}_{\mathrm{ZD}}^{-1}\mathbf{S}_{\mathrm{ZD}}^T\right)^{-1}\mathbf{S}_{\mathrm{ZD}}\mathbf{C}_{\mathrm{ZD}}^{-1}. \tag{D.2}$$

Next, the Neumann series is used to obtain a Taylor expansion for the inverse of a matrix. Recall that for a matrix $\mathbf{T}$ it holds in general

$$(1 - \mathbf{T}) \cdot (1 + \mathbf{T} + \mathbf{T}^2 + \cdots + \mathbf{T}^n) = 1 - \mathbf{T}^{n-1}. \tag{D.3}$$

Thus, in case $\|\mathbf{T}\|_2 < 1$, the Neumann series converges and it holds

$$(1 - \mathbf{T})^{-1} = \sum_{k=0}^{\infty} \mathbf{T}^k. \tag{D.4}$$

Now, for the decomposition $\mathbf{A} = \mathbf{D} - \mathbf{O}$ with $\mathbf{D}$ being diagonal, we can write

$$\mathbf{A} = \mathbf{D}^{\frac{1}{2}}\left(1 - \left(1 - \mathbf{D}^{-\frac{1}{2}}\mathbf{A}\mathbf{D}^{-\frac{1}{2}}\right)\right)\mathbf{D}^{\frac{1}{2}}$$

and for the inverse, it follows

$$\begin{aligned}
\mathbf{A}^{-1} &= \mathbf{D}^{-\frac{1}{2}}\left(1 - \left(1 - \mathbf{D}^{-\frac{1}{2}}\mathbf{A}\mathbf{D}^{-\frac{1}{2}}\right)\right)^{-1}\mathbf{D}^{-\frac{1}{2}} \\
&= \mathbf{D}^{-\frac{1}{2}}\left(1 - \mathbf{D}^{-\frac{1}{2}}\mathbf{O}\mathbf{D}^{-\frac{1}{2}}\right)^{-1}\mathbf{D}^{-\frac{1}{2}} \\
&= \mathbf{D}^{-\frac{1}{2}}\left(\sum_{k=0}^{\infty}\left(\mathbf{D}^{-\frac{1}{2}}\mathbf{O}\mathbf{D}^{-\frac{1}{2}}\right)^k\right)\mathbf{D}^{-\frac{1}{2}} \\
&= \mathbf{D}^{-1} + \mathbf{D}^{-1}\mathbf{O}\mathbf{D}^{-1} + \mathbf{D}^{-1}\mathbf{O}\mathbf{D}^{-1}\mathbf{O}\mathbf{D}^{-1} + \ldots,
\end{aligned} \tag{D.5}$$

converging as long as $\|\mathbf{D}^{-1/2}\mathbf{O}\mathbf{D}^{-1/2}\|_2 < 1$. With this expansion, we find in lowest order in the diagonal entries

$$\mathbf{C}^{-1} = \mathbf{D}_{\mathrm{ZD}}^{-1} + \mathbf{D}_{\mathrm{ZD}}^{-1}\mathbf{O}_{\mathrm{ZD}}\mathbf{D}_{\mathrm{ZD}}^{-1} + \mathbf{D}_{\mathrm{ZD}}^{-1}\mathbf{S}_{\mathrm{ZD}}^{T}\mathbf{D}_{\mathrm{R}}^{-1}\mathbf{S}_{\mathrm{ZD}}\mathbf{D}_{\mathrm{ZD}}^{-1} + \mathcal{O}\left(\mathbf{D}^{-4}\right). \tag{D.6}$$

E Proof of the transformation in Eq. 3.64

For more TLSs, it is expedient to express the rate equation $\mathcal{L}_{\mathrm{D}}$ with the help of spin creation and annihilation operators. These operators obey the anticommutation relations

$$\{\sigma_j^-, \sigma_j^+\} = 1, \quad \{\sigma_j^-, \sigma_j^-\} = \{\sigma_j^+, \sigma_j^+\} = 0 \tag{E.1}$$

for the same element, while for different elements $j \neq i$ they commute with each other:

$$[\sigma_j^-, \sigma_i^-] = [\sigma_j^+, \sigma_i^+] = [\sigma_j^-, \sigma_i^+] = 0. \tag{E.2}$$

Using these operators, the non-Hermitian rate equation $\mathcal{L}_{\mathrm{D}}$ can now be expressed for an arbitrary number of TLSs. For brevity, we will focus on the scenario that the TLSs are not interacting with each other. We have

$$\begin{aligned}
\mathcal{L}_{\mathrm{D}} = &\sum_{j=0}^{n} \Gamma_{\downarrow}^{j}(\sigma_j^- - \sigma_j^+\sigma_j^-) + \sum_{j=0}^{n} \Gamma_{\uparrow}^{j}(\sigma_j^+ - \sigma_j^-\sigma_j^+) \\
&+ \sum_{j=1}^{n} \Gamma_{\mathrm{qt}}^{\delta_j}(\sigma_j^+\sigma_0^- - \sigma_j^-\sigma_j^+\sigma_0^+\sigma_0^- \\
&\qquad + \sigma_j^-\sigma_0^+ - \sigma_j^+\sigma_j^-\sigma_0^-\sigma_0^+) \\
&+ \sum_{j=1}^{n} \Gamma_{\mathrm{qt}}^{\sigma_j}(\sigma_j^+\sigma_0^+ - \sigma_j^+\sigma_j^-\sigma_0^+\sigma_0^- \\
&\qquad + \sigma_j^-\sigma_0^- - \sigma_j^-\sigma_j^+\sigma_0^-\sigma_0^+),
\end{aligned} \tag{E.3}$$

where we use the index $j = 0$ for the qubit and $j > 0$ to denote the TLSs. The covariant vectors $\langle v_i|$ such that $p_i = \langle v_i|\,\rho_{\mathrm{D}}$ as well as $1 = \langle 1|\,\rho_{\mathrm{D}}$ can be created via

$$\langle v_i| = \langle 0|\,\sigma_i^- \prod_{j\neq i}(1 + \sigma_j^-), \tag{E.4}$$

$$\langle 1| = \langle 0| \prod_{j}(1 + \sigma_j^-). \tag{E.5}$$

The next step is to show that these vectors are mapped onto each other when they are applied to the rate equation in Eq. E.3 from the left. The first sum simply results in

$\langle v_i| \mathcal{L}_\mathrm{D} = -\Gamma_\downarrow^i \langle v_i| + \cdots$. In a similar way, the second sum vanishes for $j \neq i$. However, for $j = i$ the simplification is more intricate. Here, we can show

$$\begin{aligned}\langle 0| \sigma_i^-(\sigma_i^+ - \sigma_i^-\sigma_i^+)\cdots &= \langle 0| \sigma_i^-(\sigma_i^+ - 1 + \sigma_i^+\sigma_i^-)\cdots \\ &= -\langle v_i| + \langle 0| (1 + \sigma_i^- - 1)(\sigma_i^+ + \sigma_i^+\sigma_i^-)\cdots \\ &= -\langle v_i| + \langle 0| (1 + \sigma_i^-)(\sigma_i^+ + 1 - \sigma_i^-\sigma_i^+)\cdots - \cancel{\langle 0| (\sigma_i^+ \cdots} \\ &= -\langle v_i| + \langle 1| - \underbrace{\langle 0| (1 + \sigma_i^-)(\sigma_i^+ - \sigma_i^-\sigma_i^+)\cdots}_{\cancel{\langle 0|\sigma_i^+\cdots}} \\ &= -\langle v_i| + \langle 1| ,\end{aligned} \tag{E.6}$$

thus, $\langle v_i| \mathcal{L}_\mathrm{D} = \cdots + \Gamma_\downarrow^i(\langle 1| - \langle v_i|) + \cdots$. The third sum yields

$$\langle v_0| \mathcal{L}_\mathrm{D} = \cdots + \sum_{j=1}^{n} \Gamma_\mathrm{qt}^{\delta_j} \left(\langle v_j| - \langle v_0|\right) + \cdots , \tag{E.7}$$

$$\langle v_i| \mathcal{L}_\mathrm{D} = \cdots + \Gamma_\mathrm{qt}^{\delta_i} \left(\langle v_0| - \langle v_i|\right) + \cdots \quad \text{for } i > 0. \tag{E.8}$$

Finally, the fourth sum goes to

$$\langle v_0| \mathcal{L}_\mathrm{D} = \cdots + \sum_{j=1}^{n} \Gamma_\mathrm{qt}^{\sigma_j} \left(\langle v_j| - \langle v_0| + \langle 1|\right) , \tag{E.9}$$

$$\langle v_i| \mathcal{L}_\mathrm{D} = \cdots + \Gamma_\mathrm{qt}^{\sigma_i} \left(\langle v_0| - \langle v_i| + \langle 1|\right) \quad \text{for } i > 0. \tag{E.10}$$

The last two equations can be incorporated into the previous equations by redefining the rates as stated in the main text. Thus, we have $\langle v_i| \mathcal{L}_\mathrm{D} = \sum_j \bar{\mathbf{A}}_{ij} \langle v_j| + \bar{\Gamma}_{\uparrow,i} \langle 1|$. Consequently, for any evolution following $\dot{\rho}_\mathrm{D} = \mathcal{L}_\mathrm{D}\rho_\mathrm{D}$, we obtain the Solomon equations for the expectation values of the populations:

$$\dot{p}_i = \langle v_i| \dot{\rho}_\mathrm{D} = \langle v_i| \mathcal{L}_\mathrm{D}\rho_\mathrm{D} = \sum_j \bar{\mathbf{A}}_{ij} p_j + \bar{\Gamma}_{\uparrow,i}. \tag{E.11}$$

F Analytic solutions and approximations

F.1 The case of identical cross-relaxation rates

As discussed in the main text, the system dynamics must be governed by the two eigenvectors $\mathbf{v}_{0,2} = \begin{pmatrix} x_{0,2} & 1 \ldots 1 \end{pmatrix}^T$ and corresponding eigenvalues $\lambda_{0,2} = -\Gamma_\mathrm{qt} x_{0,2} + \Gamma_\mathrm{t} + \Gamma_\mathrm{qt}$ with

$$-2\Gamma_\mathrm{qt} x_{0,2} = \underbrace{\Gamma_\mathrm{q} + (n-1)\Gamma_\mathrm{qt} - \Gamma_\mathrm{t}}_{(\,)} \pm \underbrace{\sqrt{(\,)^2 + 4n\Gamma_\mathrm{qt}^2}}_{\sqrt{\ }} . \tag{F.1}$$

Given the initial out-of-equilibrium populations of the qubit $p^*_{q,0} := p^*_q(t = 0)$ and of the TLSs $p^*_{t,0} := p^*_t(t = 0)$, it holds

$$\mathbf{p}^*(0) = \frac{p^*_{t,0} - q}{2}\mathbf{v}_0 + \frac{p^*_{t,0} + q}{2}\mathbf{v}_2 \tag{F.2}$$

with $q = (2\Gamma_{qt}p^*_{q,0} + (\,)\,p^*_{t,0})/\sqrt{}$, and the qubit and TLS dynamics finally read

$$p^*_q(t) = \frac{x_0}{2}(p^*_{t,0} - q)e^{-\lambda_0 t} + \frac{x_2}{2}(p^*_{t,0} + q)e^{-\lambda_2 t}, \tag{F.3}$$

$$p^*_t(t) = \frac{1}{2}(p^*_{t,0} - q)e^{-\lambda_0 t} + \frac{1}{2}(p^*_{t,0} + q)e^{-\lambda_2 t}. \tag{F.4}$$

The time-dependent transition rates and the equilibrium population can now be easily computed using Eqs. 3.67, 3.68 and 3.70.

F.2 The case of distributed cross-relaxation rates

The Pick function of the distribution Eq. 3.81 may first be brought into the form

$$f(\lambda) = \Gamma_q - \Gamma_t - \frac{a}{a/\lambda'} - \sum_{k=1}^{\infty} \frac{a}{a/\lambda' - k^2} \tag{F.5}$$

with $\lambda' = \lambda - \Gamma_t$. Here, we see the nature of the Pick function: it is a meromorphic function of z in the complex plane, with $z = a/\lambda' = a/(\lambda - \Gamma_t)$. The sum in Eq. F.5 can be expressed in closed form. It holds

$$\frac{f(z)}{a} = \gamma - \frac{1}{2z} - \frac{\pi}{2\sqrt{z}}\cot \pi\sqrt{z}, \tag{F.6}$$

where $\gamma = (\Gamma_q - \Gamma_t)/a$ was introduced. The normalization can now conveniently be calculated via

$$\begin{aligned} \|v_m\|^2 &= -f'(\lambda)\Big|_{\lambda=\lambda_m} = z^2 \frac{\partial}{\partial z}\frac{f(z)}{a}\Big|_{z=z_m} \\ &= \frac{1}{2} + \left(\frac{\pi^2}{4} - \frac{\gamma}{2}\right) z_m + \gamma^2 z_m^2. \end{aligned} \tag{F.7}$$

Here, we see that $\|v_m\|^2$ is a continuous function of z_m, which allows the use of simple approximations for the roots z_m in the next step.

So far, our analysis is still exact. We are left with the evaluation of the sum in Eq. 3.78, which becomes an integral on long time scales. This motivates to interpret z as a continuous function of m. The structure of Eq. 3.72 with the rates Γ^k_{qt}, given by Eq. 3.81, suggests

$$\lambda_m = \frac{a}{(m + \frac{1}{2} + \delta(m))^2} + \Gamma_{\mathrm{t}} \tag{F.8}$$

$$\Rightarrow \quad z(m) = \left(m + \frac{1}{2} + \delta(m)\right)^2, \tag{F.9}$$

where we defined the deviation $\delta(m) \in (-1/2, 1/2)$.

As a side note, a very good approximation can be found for the mth-root when the Pick function is approximated piecewise by its surrounding poles, as shown in Fig. 3.9. This approximation will be used later for the analysis of a more general rate distribution. Here, we obtain:

$$z(m) \approx m^2 + m + 1/\gamma + 1/2 - \mathrm{sgn}(\gamma)\sqrt{m^2 + m + 1/\gamma^2 + 1/4}, \tag{F.10}$$

which becomes $z(m) = m^2 + m + 1/2$ for $\gamma = 0$. At first glance, when the derivative f' is also approximated by the surrounding poles, insertion of Eq. F.10 gives the correct limit $\|v_m\|^2 \to \gamma^2 m^4$ for large m and $\gamma^2 > 0$. However, when checking $\gamma^2 = 0$, one obtains the slightly inaccurate limit $\|v_m\|^2 \to 2m^2$ instead of $\|v_m\|^2 \to \beta m^2 = (\pi^2/4)m^2$ according to Eq. F.7.

With the function $z(m)$ we have everything needed to evaluate the sum in Eq. 3.78 on long time scales. We begin with the simpler scenario where $\gamma^2 = 0$, which essentially describes the qubit relaxing into the TLS environment. Defining $y(m) = \left(x(m) + \frac{1/2+\delta(m)}{\sqrt{at}}\right)$ and $x(m) = \frac{m}{\sqrt{at}}$, we have

$$\begin{aligned}
\sum_{m=0}^{\infty} \frac{e^{-\lambda'_m t}}{\|v_m\|^2} &= \frac{1}{\sqrt{at}} \sum_{m=0}^{\infty} \frac{e^{-\frac{1}{y(m)^2}}}{\frac{1/2}{at} + \beta y(m)} \frac{1}{\sqrt{at}} \\
&\approx \frac{1}{\sqrt{at}} \int_0^{\infty} \frac{e^{-\frac{1}{y(m(x))^2}}}{\frac{1/2}{at} + \beta y(m(x))} \mathrm{d}x \approx \frac{1}{\sqrt{at}} \int_0^{\infty} \frac{e^{-\frac{1}{x^2}}}{\frac{1/2}{at} + \beta x^2} \mathrm{d}x \\
&\approx \frac{1}{\sqrt{at}} \int_0^{\infty} \frac{e^{-\frac{1}{x^2}}}{\beta x^2} \mathrm{d}x = \frac{\sqrt{\pi}}{2\beta} \frac{1}{(at)^{1/2}},
\end{aligned} \tag{F.11}$$

where at first the sum was approximated by an integral, requiring $\sqrt{at} \gg 1$, which also allows the second approximation $(1/2 + \delta)/\sqrt{at} \approx 0$, since $|\delta| < 1/2$. In the last step, the denominator was approximated, which is valid when $\sqrt{at} \gg 1/\sqrt{2\beta}$.

The scenario $\gamma^2 > 0$ can be treated similarly. We have

$$\sum_{m=0}^{\infty} \frac{e^{-\lambda'_m t}}{\|v_m\|^2} \approx \frac{1}{(at)^{3/2}} \int_0^\infty \frac{e^{-\frac{1}{x^2}}}{\frac{1/2}{(at)^2} + \frac{\beta}{at}x^2 + \gamma^2 x^4} \frac{1}{\sqrt{at}}$$
$$\approx \frac{1}{(at)^{3/2}} \int_0^\infty \frac{e^{-\frac{1}{x^2}}}{\gamma^2 x^4} \mathrm{d}x = \frac{\sqrt{\pi}}{4\gamma^2} \frac{1}{(at)^{3/2}}, \tag{F.12}$$

which in addition to the previous scenario requires $\sqrt{at} \gg \sqrt{\beta/\gamma^2}$. The final limit relaxation behaviors are stated in the main text. The approximation of the sum by the integral is depicted in Fig. F.1.

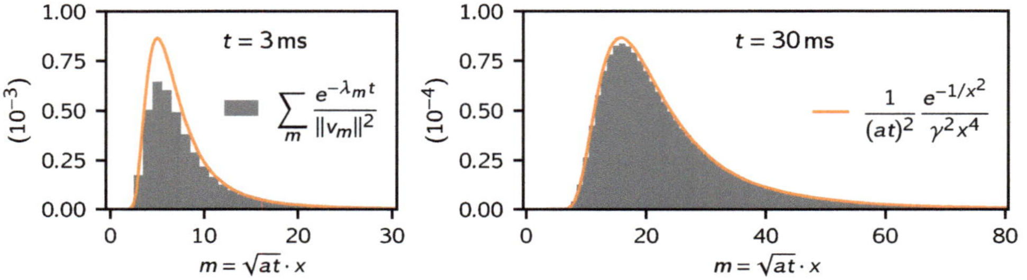

Figure F.1: Evaluation of the sum. The parameters are chosen similarly to the experiment with $a = 20\,\mathrm{kHz}$, $\Gamma_\mathrm{q} = 10\,\mathrm{kHz}$ and $\Gamma_\mathrm{t} = 0$ and correspond to the Pick function in Fig. 3.9.

Generalization

For the generalized distribution Eq. 3.87, we find analogously

$$f(\lambda) = \Gamma_\mathrm{q} - \Gamma_\mathrm{t} - \frac{a}{a/\lambda'} - \sum_{k=1}^{\infty} \frac{a}{a/\lambda' - k^d}. \tag{F.13}$$

Unfortunately, to our knowledge, this function can not be expressed in a closed form, nor can its derivative be expressed via the function itself. For integer values of d, one can express f as a sum of digamma functions, which for even integers can be rewritten as a sum of cotangents. In App. F.2, we show the relevant cases $d = 3$ and $d = 4$. We will therefore proceed as discussed before and approximate f piecewise by its surrounding poles. For $\gamma^2 > 0$, we then find $\|v_m\|^2 \to \gamma^2(m + 1/2 + \delta(m))^{2d}$ for large m in congruence with Eq. F.7. For $\gamma^2 = 0$ one finds in leading order $\|v_m\|^2 \to 8/d^2 m^2$, which surprisingly is always quadratic in m and not with the power of d, as one might surmise from Eq. F.7. Note, as discussed before, the prefactor $8/d^2$ is only an approximation. For instance, for $d = 3$ the correct prefactor is $4\pi^2/27$ and for $d = 4$

one finds $\pi^2/8$ (cf. App. F.2). We therefore continue by reintroducing the prefactor β via $||v_m||^2 \to \beta m^2$ for large m. Similar as previously, we have for $\gamma^2 = 0$

$$\begin{aligned}\sum_{m=0}^{\infty} \frac{e^{-\lambda'_m t}}{||v_m||^2} &\approx \frac{1}{(at)^{1/d}} \sum_{m=0}^{\infty} \frac{e^{-\frac{1}{y(m)^d}}}{\beta \frac{m^2}{(at)^{2/d}}} \frac{1}{(at)^{1/d}} \\ &\approx \frac{1}{(at)^{1/d}} \int_0^{\infty} \frac{e^{-\frac{1}{x^d}}}{\beta x^2} \mathrm{d}x = \frac{\Gamma(1+\frac{1}{d})}{\beta} \frac{1}{(at)^{1/d}}\end{aligned} \tag{F.14}$$

with the approximation becoming valid for $(at)^{1/d} \gg 1$. Here, Γ denotes the Gamma function. For $\gamma^2 > 0$ we find

$$\begin{aligned}\sum_{m=0}^{\infty} \frac{e^{-\lambda'_m t}}{||v_m||^2} &\approx \frac{1}{(at)^{2-1/d}} \sum_{m=0}^{\infty} \frac{e^{-\frac{1}{y(m)^d}}}{\gamma^2 \frac{y^{2d}}{(at)^2}} \frac{1}{(at)^{1/d}} \\ &\approx \frac{1}{(at)^{2-1/d}} \int_0^{\infty} \frac{e^{-\frac{1}{x^d}}}{\gamma^2 x^{2d}} \mathrm{d}x = \frac{\Gamma(2-\frac{1}{d})}{\gamma^2 d} \frac{1}{(at)^{2-1/d}},\end{aligned} \tag{F.15}$$

which is again only valid when $(at)^{1/d} \gg 1$. The final relaxation behaviors on long time scales are stated in the main text.

Special solutions

In the following, we will derive a few more exact solutions for the Pick function. We begin with the Lorentzian distribution as defined in Eq. 3.80. Note that in the case of $2bc \in \mathbb{Z}$, the cross-relaxation rates are not distinct from each other, and the Pick function is not directly applicable. In this case, the irrelevant eigenvalues given by the degenerate cross-relaxation rates have to be treated beforehand. The analysis is then very similar to the one shown in the following, which is valid for $2bc \notin \mathbb{Z}$. The Pick function can be rewritten as

$$\begin{aligned}\frac{f(\lambda)}{a} &\equiv \gamma - \frac{1}{z} - \sum_{h=-\infty}^{\infty} \frac{b^2}{(z-1)b^2 - (k+bc)^2} \\ &= \gamma - \frac{1}{z} - \frac{\pi b}{2\sqrt{z-1}} \left[\cot\left(\pi bc + \pi b\sqrt{z-1}\right) - \cot\left(\pi bc - \pi b\sqrt{z-1}\right) \right].\end{aligned} \tag{F.16}$$

Moving from the first to the second line requires $z > 1$, which becomes valid for $\lambda \gtrsim \lambda_0$. Unfortunately, when taking the derivative, the divergent cotangents terms can not be removed simultaneously, except for the symmetric cases $bc \in \{0, 1/4, 1/2\}$. For instance, for $bc = 1/4$ we obtain

$$\frac{f(\lambda)}{a} = \gamma - \frac{1}{z} + \frac{\pi b}{\sqrt{z-1}} \tan\left(2\pi b\sqrt{z-1}\right) \tag{F.17}$$

and

$$\|v_m\|^2 = 1 + \frac{1}{2}\frac{(\gamma z_m - 1)z_m}{(z_m - 1)^2} + \frac{\pi^2 b^2 z_m^2}{z_m - 1} + (\gamma z_m - 1)^2. \tag{F.18}$$

Note, when expressing this solution as a continuous function $z(m) = (m/2 + 1/4 + \delta(m)/2)^2 = (m + 1/2 + \delta(m))^2/4$, the TLSs with negative and positive detuning have to be considered. In order to compare this result with the one from the distribution defined in Eq. 3.81, the parameter a must be scaled by a factor of 4.

Next, we discuss the distribution defined in Eq. 3.87, which can be solved analytically for integer values $d > 1$. We only present the illustrative calculations of $d = 3$ and $d = 4$. For $d = 4$, it holds

$$\begin{aligned}
\frac{f(\lambda)}{a} &= \gamma - \frac{1}{z} - \sum_{k=1}^{\infty} \frac{1}{z - k^4} \\
&= \gamma - \frac{1}{z} - \frac{1}{4z^{3/4}} \sum_{k=1}^{\infty} \frac{1}{\sqrt[4]{z} + k} + \frac{1}{\sqrt[4]{z} - k} + \frac{1}{\sqrt[4]{z} + ik} + \frac{1}{\sqrt[4]{z} - ik} \\
&= \gamma - \frac{1}{2z} - \frac{\pi \cot\left(\pi\sqrt[4]{z}\right)}{4z^{3/4}} - \frac{\pi \coth\left(\pi\sqrt[4]{z}\right)}{4z^{3/4}}.
\end{aligned} \tag{F.19}$$

Expressing the oscillating cotangent through regular terms via the Pick function yields

$$\begin{aligned}
\|v_m\|^2 = \frac{3}{8} + \frac{\pi z_m^{1/4}}{4}\coth\left(\pi\sqrt[4]{z_m}\right) + \frac{\pi^2 z_m^{2/4}}{8} - \frac{\gamma z_m}{4} \\
- \frac{\pi \gamma z_m^{5/4}}{2}\coth\left(\pi\sqrt[4]{z_m}\right) + \gamma^2 z_m^2.
\end{aligned} \tag{F.20}$$

Representing this solution again as a continuous function $z(m) = (m + 1/2 + \delta(m))^4$, we obtain the expected behavior in the limit $m \to \infty$, and we find $\beta = \pi^2/8$. In general, when d is even, f contains $d/2$ cotangent functions with complex coefficients. Rewriting the oscillating cotangent on the real axis using the Pick function gives a continuous function for $\|v_m\|^2$.

In principle, we can use the same strategy for $d = 3$:

$$
\begin{aligned}
\frac{f(\lambda)}{a} &= \gamma - \frac{1}{z} - \sum_{k=1}^{\infty} \frac{1}{z - k^3} \\
&= \cdots - \frac{1}{3z^{2/3}} \sum_{k=1}^{\infty} \frac{1}{\sqrt[3]{z} - k} + \frac{e^{\frac{2\pi i}{3}}}{\sqrt[3]{z}e^{\frac{2\pi i}{3}} - k} + \frac{e^{-\frac{2\pi i}{3}}}{\sqrt[3]{z}e^{-\frac{2\pi i}{3}} - k} \\
&= \cdots - \frac{1}{3z^{2/3}} \Big[\psi_0\left(1 - \sqrt[3]{z}\right) + e^{\frac{2\pi i}{3}} \psi_0\left(1 - \sqrt[3]{z}e^{\frac{2\pi i}{3}}\right) \\
&\qquad + e^{-\frac{2\pi i}{3}} \psi_0\left(1 - \sqrt[3]{z}e^{-\frac{2\pi i}{3}}\right)\Big] \\
&= \cdots - \frac{1}{3z^{2/3}} \Big[\pi \cot\left(\pi\sqrt[3]{z}\right) + \psi_0(\sqrt[3]{z}) \\
&\qquad - e^{\frac{2\pi i}{6}} \psi_0(\sqrt[3]{z}e^{\frac{2\pi i}{6}}) - e^{-\frac{2\pi i}{6}} \psi_0(\sqrt[3]{z}e^{-\frac{2\pi i}{6}}) - \frac{2}{\sqrt[3]{z}}\Big] \\
&= \gamma - \frac{1}{3z} - \frac{1}{3z^{2/3}} \left[\pi \cot\left(\pi\sqrt[3]{z}\right) + \frac{\pi}{\sqrt{3}} + \frac{1}{2z^{1/3}} + I(\sqrt[3]{z})\right] \\
&= \gamma - \frac{1}{2z} - \frac{\pi \cot(\pi\sqrt[3]{z})}{3z^{2/3}} + \frac{\pi}{\sqrt{27}z^{2/3}} + \frac{I(\sqrt[3]{z})}{3z^{2/3}},
\end{aligned}
\tag{F.21}
$$

where ψ_0 is the digamma function and $I(\sqrt[3]{z}) = \mathcal{O}(1/z^{4/3})$ is an integral expression that vanishes continuously and sufficiently fast as $z \longrightarrow \infty$. The integral expression enters as

$$
\psi(z) = \ln(z) - \frac{1}{2z} - \int_0^{\infty} \left(\frac{1}{2} - \frac{1}{t} + \frac{1}{e^{-t} - 1}\right) e^{-zt} \mathrm{d}t \tag{F.22}
$$

for $\mathrm{Re}(z) > 0$. By taking the derivative and removing the divergent cotangent, one obtains $\beta = 4\pi^2/27$. The procedure essentially works for any integer $d \geq 2$. In case one is only interested in the closed form expression for $\|v_m\|^2$, one may only use the reflection formula once to yield the cotangent via $\psi_0(1 - \sqrt[d]{z}) = \pi \cot(\pi\sqrt[d]{z}) + \psi_0(\sqrt[d]{z})$. Here, the poles on the negative real axis are canceled with $\psi_0(\sqrt[d]{z})$.

G Detailed fabrication steps

Table G.1: Fabrication steps for the three-angle fabrication of granular aluminum fluxoniums.

#	description	parameters
1	EL-13 spin coating	ramp rate: 1000 rpm/s, speed: 2000 rpm, spin time: 100 s, HP: 200 °C, 5 min → ~ 950 nm
2	A-4 spin coating	ramp rate: 1000 rpm/s, speed: 2000 rpm, spin time: 100 s, HP: 200 °C, 5 min → ~ 200 nm
3	gold sputtering	
4	e-beam writing (JOEL 50 keV)	
5	gold removal	Lugol solution 15 %, 10 s
6	development	IPA / H_2O 3:1 @ 6 °C, 90 s, slightly move wafer
7	e-beam evaporation (PASSYS)	
	1. plasma descum	O_2 10 sccm, Ar 5 sccm, 2 min, voltage: 200 V, current: 10 mA
	2. getter pumping	titanium evap. with shutter closed, 0.2 nm/s, 2 min
	3. Al evaporation	1.0 nm/s, tilt: 30°, 20 nm
	4. junction oxidation	15 mbar, 5:30 min • ~ 40 s needed to reach pressure
	5. Al evaporation	1.0 nm/s, tilt: −30°, 30 nm
	• later added:	
	5.1 argon milling	Ar 4 sccm, 1 min, voltage: 400 V, current: 15 mA
	6. grAl evaporation	0.3 nm/s, O_2 1.0-1.5 sccm, tilt: 0°, 40 nm • later changed to 1.0 nm/s, O_2 4.0-6.0 sccm • further improvement: (i) rate stabilization to 0.8 nm/s with no oxygen (ii) add oyxgen and adjust rate control to 1.0 nm/s (iii) rate stabilization for 10 s, open shutter
8	lift-off	N-Ethyl-2-pyrrolidone (NEP) @ 90 °C, 90 min, rinsing in IPA and Ethanol

Bibliography

[1] V. V. Sivak, A. Eickbusch, B. Royer, S. Singh, I. Tsioutsios, et al., "Real-time quantum error correction beyond break-even", Nature **616**, 50–55 (2023) (cit. on p. 5).

[2] A. A. Clerk, K. W. Lehnert, P. Bertet, J. R. Petta, and Y. Nakamura, "Hybrid quantum systems with circuit quantum electrodynamics", Nat. Phys. **16**, 257–267 (2020) (cit. on p. 5).

[3] Z. Wang, L. Balembois, M. Rančić, E. Billaud, M. Le Dantec, et al., "Single-electron spin resonance detection by microwave photon counting", Nature **619**, 276–281 (2023) (cit. on pp. 5, 95, 105).

[4] S. Léger, J. Puertas-Martínez, K. Bharadwaj, R. Dassonneville, J. Delaforce, et al., "Observation of quantum many-body effects due to zero point fluctuations in superconducting circuits", Nat. Commun. **10**, 1–8 (2019) (cit. on p. 5).

[5] Q. Guo, C. Cheng, Z.-H. Sun, Z. Song, H. Li, et al., "Observation of energy-resolved many-body localization", Nat. Phys. **17**, 234–239 (2021) (cit. on p. 5).

[6] A. Ronzani, B. Karimi, J. Senior, Y.-C. Chang, J. T. Peltonen, C. Chen, and J. P. Pekola, "Tunable photonic heat transport in a quantum heat valve", Nat. Phys. **14**, 991–995 (2018) (cit. on p. 5).

[7] J. Stevens, D. Szombati, M. Maffei, C. Elouard, R. Assouly, et al., "Energetics of a Single Qubit Gate", Phys. Rev. Lett. **129**, 110601 (2022) (cit. on pp. 5, 6, 77, 105, 137).

[8] P. Campagne-Ibarcq, P. Six, L. Bretheau, A. Sarlette, M. Mirrahimi, P. Rouchon, and B. Huard, "Observing Quantum State Diffusion by Heterodyne Detection of Fluorescence", Phys. Rev. X **6**, 011002 (2016) (cit. on p. 5).

[9] Z. K. Minev, S. O. Mundhada, S. Shankar, P. Reinhold, R. Gutiérrez-Jáuregui, R. J. Schoelkopf, M. Mirrahimi, H. J. Carmichael, and M. H. Devoret, "To catch and reverse a quantum jump mid-flight", Nature **570**, 200–204 (2019), (cit. on pp. 5, 79, 105).

[10] E. Blumenthal, C. Mor, A. A. Diringer, L. S. Martin, P. Lewalle, D. Burgarth, K. B. Whaley, and S. Hacohen-Gourgy, "Demonstration of universal control between non-interacting qubits using the Quantum Zeno effect", npj Quantum Inf. **8**, 1–6 (2022) (cit. on p. 5).

[11] O. V. Ivakhnenko, S. N. Shevchenko, and F. Nori, "Simulating quantum dynamical phenomena using classical oscillators: Landau-Zener-Stückelberg-Majorana interferometry, latching modulation, and motional averaging", Sci. Rep. **8**, 1–11 (2018) (cit. on p. 5).

[12] V. V. Shende, S. S. Bullock, and I. L. Markov, "Synthesis of quantum-logic circuits", IEEE Trans. Comput. Aided Des. Integr. Circuits Syst. **25**, 1000–1010 (2006) (cit. on p. 6).

[13] A. Montanaro, "Quantum algorithms: an overview", npj Quantum Inf. **2**, 1–8 (2016) (cit. on p. 6).

[14] A. Bassi, K. Lochan, S. Satin, T. P. Singh, and H. Ulbricht, "Models of wave-function collapse, underlying theories, and experimental tests", Rev. Mod. Phys. **85**, 471–527 (2013) (cit. on pp. 6, 47).

[15] V. E. Manucharyan, J. Koch, L. I. Glazman, and M. H. Devoret, "Fluxonium: Single Cooper-Pair Circuit Free of Charge Offsets", Science **326**, 113–116 (2009) (cit. on pp. 6, 17, 113, 127).

[16] G. Deutscher, M. Gershenson, E. Grünbaum, and Y. Imry, "Granular Superconducting Films", J. Vac. Sci. Technol. **10**, 697–701 (1973) (cit. on pp. 6, 117, 119).

[17] H. Rotzinger, S. T. Skacel, M. Pfirrmann, J. N. Voss, J. Münzberg, S. Probst, P. Bushev, M. P. Weides, A. V. Ustinov, and J. E. Mooij, "Aluminium-oxide wires for superconducting high kinetic inductance circuits", Supercond. Sci. Technol. **30**, 025002 (2016) (cit. on pp. 6, 102, 103, 117, 119, 154).

[18] L. Grünhaupt, M. Spiecker, D. Gusenkova, N. Maleeva, S. T. Skacel, et al., "Granular aluminium as a superconducting material for high-impedance quantum circuits", Nat. Mater. **18**, 816–819 (2019) (cit. on pp. 6, 17, 20, 23, 118, 125, 154).

[19] D. Rieger, S. Günzler, M. Spiecker, P. Paluch, P. Winkel, L. Hahn, J. K. Hohmann, A. Bacher, W. Wernsdorfer, and I. M. Pop, "Granular aluminium nanojunction fluxonium qubit", Nat. Mater. **22**, 194–199 (2023) (cit. on pp. 7, 113).

[20] P. Winkel, K. Borisov, L. Grünhaupt, D. Rieger, M. Spiecker, F. Valenti, A. V. Ustinov, W. Wernsdorfer, and I. M. Pop, "Implementation of a Transmon Qubit Using Superconducting Granular Aluminum", Phys. Rev. X **10**, 031032 (2020) (cit. on pp. 7, 118, 119).

[21] M. Spiecker, P. Paluch, N. Gosling, N. Drucker, S. Matityahu, et al., "Two-level system hyperpolarization using a quantum Szilard engine", Nat. Phys. **19**, 1320–1325 (2023) (cit. on pp. 9, 11, 14, 19–21, 24, 26–33, 35, 37, 38, 40, 43, 48, 60, 70, 87, 157, 161).

[22] L. Szilárd, "Über die Entropieverminderung in einem thermodynamischen System bei Eingriffen intelligenter Wesen", Z. Phys. **53**, 840–856 (1929) (cit. on pp. 9, 11, 14, 39).

[23] R. Landauer, "Irreversibility and Heat Generation in the Computing Process", IBM J. Res. Dev. **5**, 183–191 (1961) (cit. on p. 10).

[24] A. Bérut, A. Arakelyan, A. Petrosyan, S. Ciliberto, R. Dillenschneider, and E. Lutz, "Experimental verification of Landauer's principle linking information and thermodynamics", Nature **483**, 187–189 (2012) (cit. on p. 10).

[25] R. Gaudenzi, E. Burzurí, S. Maegawa, H. S. J. van der Zant, and F. Luis, "Quantum Landauer erasure with a molecular nanomagnet", Nat. Phys. **14**, 565–568 (2018) (cit. on p. 10).

[26] C. H. Bennett, "Notes on Landauer's principle, reversible computation, and Maxwell's Demon", Studies in History and Philosophy of Science Part B: Studies in History and Philosophy of Modern Physics **34**, 501–510 (2003) (cit. on p. 10).

[27] W. H. Zurek, "Maxwell's Demon, Szilard's Engine and Quantum Measurements", Frontiers of Nonequilibrium Statistical Physics (Springer, Boston, MA, Boston, MA, USA, 1986), pp. 151–161, (cit. on p. 10).

[28] A. Aydin, A. Sisman, and R. Kosloff, "Landauer's Principle in a Quantum Szilard Engine without Maxwell's Demon", Entropy **22**, 294 (2020) (cit. on p. 10).

[29] P. A. Camati, J. P. S. Peterson, T. B. Batalhão, K. Micadei, A. M. Souza, R. S. Sarthour, I. S. Oliveira, and R. M. Serra, "Experimental Rectification of Entropy Production by Maxwell's Demon in a Quantum System", Phys. Rev. Lett. **117**, 240502 (2016) (cit. on p. 11).

[30] N. Cottet, S. Jezouin, L. Bretheau, P. Campagne-Ibarcq, Q. Ficheux, J. Anders, A. Auffèves, R. Azouit, P. Rouchon, and B. Huard, "Observing a quantum Maxwell demon at work", Proc. Natl. Acad. Sci. U.S.A. **114**, 7561–7564 (2017) (cit. on p. 11).

[31] J. P. S. Peterson, R. S. Sarthour, and R. Laflamme, "Implementation of a Quantum Engine Fuelled by Information", (2020), DOI: `10.48550/arXiv.2006.10136` (cit. on p. 11).

[32] J. Bengtsson, M. N. Tengstrand, A. Wacker, P. Samuelsson, M. Ueda, H. Linke, and S. M. Reimann, "Quantum Szilard Engine with Attractively Interacting Bosons", Phys. Rev. Lett. **120**, 100601 (2018) (cit. on p. 13).

[33] D. Gusenkova, M. Spiecker, R. Gebauer, M. Willsch, D. Willsch, et al., "Quantum Nondemolition Dispersive Readout of a Superconducting Artificial Atom Using Large Photon Numbers", Phys. Rev. Appl. **15**, 064030 (2021) (cit. on pp. 17, 19, 28, 125, 137, 148).

[34] A. Blais, R.-S. Huang, A. Wallraff, S. M. Girvin, and R. J. Schoelkopf, "Cavity quantum electrodynamics for superconducting electrical circuits: An architecture for quantum computation", Phys. Rev. A **69**, 062320 (2004) (cit. on pp. 18, 137, 138, 145).

[35] A. Wallraff, D. I. Schuster, A. Blais, L. Frunzio, R. S. Huang, J. Majer, S. Kumar, S. M. Girvin, and R. J. Schoelkopf, "Strong coupling of a single photon to a superconducting qubit using circuit quantum electrodynamics", Nature **431**, 162–167 (2004) (cit. on pp. 18, 137, 138, 145).

[36] M. Boissonneault, J. M. Gambetta, and A. Blais, "Dispersive regime of circuit QED: Photon-dependent qubit dephasing and relaxation rates", Phys. Rev. A **79**, 013819 (2009) (cit. on p. 18).

[37] T. Walter, P. Kurpiers, S. Gasparinetti, P. Magnard, A. Potočnik, et al., "Rapid High-Fidelity Single-Shot Dispersive Readout of Superconducting Qubits", Phys. Rev. Appl. **7**, 054020 (2017) (cit. on p. 18).

[38] M. Malekakhlagh, A. Petrescu, and H. E. Türeci, "Lifetime renormalization of weakly anharmonic superconducting qubits. I. Role of number nonconserving terms", Phys. Rev. B **101**, 134509 (2020) (cit. on p. 18).

[39] A. Petrescu, M. Malekakhlagh, and H. E. Türeci, "Lifetime renormalization of driven weakly anharmonic superconducting qubits. II. The readout problem", Phys. Rev. B **101**, 134510 (2020) (cit. on p. 18).

[40] R. Shillito, A. Petrescu, J. Cohen, J. Beall, M. Hauru, M. Ganahl, A. G. M. Lewis, G. Vidal, and A. Blais, "Dynamics of Transmon Ionization", Phys. Rev. Appl. **18**, 034031 (2022) (cit. on pp. 18, 144–146).

[41] E. A. Tholén, A. Ergül, E. M. Doherty, F. M. Weber, F. Grégis, and D. B. Haviland, "Nonlinearities and parametric amplification in superconducting coplanar waveguide resonators", Appl. Phys. Lett. **90**, 253509 (2007) (cit. on p. 18).

[42] T. Yamamoto, K. Inomata, M. Watanabe, K. Matsuba, T. Miyazaki, W. D. Oliver, Y. Nakamura, and J. S. Tsai, "Flux-driven Josephson parametric amplifier", Appl. Phys. Lett. **93**, 042510 (2008) (cit. on p. 18).

[43] C. Eichler, Y. Salathe, J. Mlynek, S. Schmidt, and A. Wallraff, "Quantum-Limited Amplification and Entanglement in Coupled Nonlinear Resonators", Phys. Rev. Lett. **113**, 110502 (2014) (cit. on p. 18).

[44] P. Winkel, I. Takmakov, D. Rieger, L. Planat, W. Hasch-Guichard, et al., "Non-degenerate parametric amplifiers based on dispersion-engineered Josephson-junction arrays", Phys. Rev. Appl. **13**, 024015 (2020) (cit. on pp. 18, 162).

[45] D. T. McClure, H. Paik, L. S. Bishop, M. Steffen, J. M. Chow, and J. M. Gambetta, "Rapid Driven Reset of a Qubit Readout Resonator", Phys. Rev. Appl. **5**, 011001 (2016) (cit. on pp. 19, 151).

[46] C. C. Bultink, M. A. Rol, T. E. O'Brien, X. Fu, B. C. S. Dikken, et al., "Active Resonator Reset in the Nonlinear Dispersive Regime of Circuit QED", Phys. Rev. Appl. **6**, 034008 (2016) (cit. on pp. 19, 151).

[47] N. Samkharadze, A. Bruno, P. Scarlino, G. Zheng, D. P. DiVincenzo, L. DiCarlo, and L. M. K. Vandersypen, "High-kinetic-inductance superconducting nanowire resonators for circuit QED in a magnetic field", Phys. Rev. Appl. **5**, 044004 (2016) (cit. on pp. 21, 22).

[48] K. Borisov, D. Rieger, P. Winkel, F. Henriques, F. Valenti, et al., "Superconducting granular aluminum resonators resilient to magnetic fields up to 1 Tesla", Appl. Phys. Lett. **117**, 120502 (2020) (cit. on pp. 21, 22, 93).

[49] C. M. Quintana, Y. Chen, D. Sank, A. G. Petukhov, T. C. White, et al., "Observation of Classical-Quantum Crossover of $1/f$ Flux Noise and Its Paramagnetic Temperature Dependence", Phys. Rev. Lett. **118**, 057702 (2017) (cit. on pp. 21, 22).

[50] M. D. Reed, B. R. Johnson, A. A. Houck, L. DiCarlo, J. M. Chow, D. I. Schuster, L. Frunzio, and R. J. Schoelkopf, "Fast reset and suppressing spontaneous emission of a superconducting qubit", Appl. Phys. Lett. **96**, 203110 (2010) (cit. on pp. 21, 157).

[51] U. von Lüpke, Y. Yang, M. Bild, L. Michaud, M. Fadel, and Y. Chu, "Parity measurement in the strong dispersive regime of circuit quantum acoustodynamics", Nat. Phys. **18**, 794–799 (2022) (cit. on p. 21).

[52] F. Yang, T. Gozlinski, T. Storbeck, L. Grünhaupt, I. M. Pop, and W. Wulfhekel, "Microscopic charging and in-gap states in superconducting granular aluminum", Phys. Rev. B **102**, 104502 (2020) (cit. on pp. 21, 39, 119).

[53] U. Vool, I. M. Pop, K. Sliwa, B. Abdo, C. Wang, et al., "Non-Poissonian Quantum Jumps of a Fluxonium Qubit due to Quasiparticle Excitations", Phys. Rev. Lett. **113**, 247001 (2014) (cit. on pp. 21, 27, 89, 90, 94).

[54] I. Nsanzineza and B. L. T. Plourde, "Trapping a single vortex and reducing quasiparticles in a superconducting resonator", Phys. Rev. Lett. **113**, 117002 (2014) (cit. on p. 21).

[55] L. Glazman and G. Catelani, "Bogoliubov quasiparticles in superconducting qubits", SciPost Phys. Lect. Notes, 031 (2021) (cit. on pp. 21, 23, 126, 153).

[56] P. Kumar, S. Sendelbach, M. A. Beck, J. W. Freeland, Z. Wang, H. Wang, C. C. Yu, R. Q. Wu, D. P. Pappas, and R. McDermott, "Origin and reduction of $1/f$ magnetic flux noise in superconducting devices", Phys. Rev. Appl. **6**, 041001 (2016) (cit. on p. 21).

[57] C. Müller, J. H. Cole, and J. Lisenfeld, "Towards understanding two-level-systems in amorphous solids: insights from quantum circuits", Rep. Prog. Phys. **82**, 124501 (2019) (cit. on pp. 21, 22).

[58] T. Astner, J. Gugler, A. Angerer, S. Wald, S. Putz, et al., "Solid-state electron spin lifetime limited by phononic vacuum modes", Nat. Mater. **17**, 313–317 (2018) (cit. on p. 21).

[59] A. Morello, J. J. Pla, P. Bertet, and D. N. Jamieson, "Donor Spins in Silicon for Quantum Technologies", Adv. Quantum Technol. **3**, 2000005 (2020) (cit. on p. 21).

[60] K. Serniak, M. Hays, G. de Lange, S. Diamond, S. Shankar, L. D. Burkhart, L. Frunzio, M. Houzet, and M. H. Devoret, "Hot nonequilibrium quasiparticles in transmon qubits", Phys. Rev. Lett. **121**, 157701 (2018) (cit. on p. 22).

[61] S. E. de Graaf, A. A. Adamyan, T. Lindström, D. Erts, S. E. Kubatkin, A. Y. Tzalenchuk, and A. V. Danilov, "Direct Identification of Dilute Surface Spins on Al_2O_3: Origin of Flux Noise in Quantum Circuits", Phys. Rev. Lett. **118**, 057703 (2017) (cit. on pp. 22, 69).

[62] C. Wang, Y. Y. Gao, I. M. Pop, U. Vool, C. Axline, et al., "Measurement and control of quasiparticle dynamics in a superconducting qubit", Nat. Commun. **5**, 1–7 (2014) (cit. on pp. 22, 23, 105).

[63] L. Grünhaupt, N. Maleeva, S. T. Skacel, M. Calvo, F. Levy-Bertrand, A. V. Ustinov, H. Rotzinger, A. Monfardini, G. Catelani, and I. M. Pop, "Loss Mechanisms and Quasiparticle Dynamics in Superconducting Microwave Resonators Made of Thin-Film Granular Aluminum", Phys. Rev. Lett. **121**, 117001 (2018) (cit. on pp. 22, 119, 153).

[64] J. Lisenfeld, A. Bilmes, S. Matityahu, S. Zanker, M. Marthaler, M. Schechter, G. Schön, A. Shnirman, G. Weiss, and A. V. Ustinov, "Decoherence spectroscopy with individual two-level tunneling defects", Sci. Rep. **6**, 1–8 (2016) (cit. on pp. 22, 39).

[65] J. Lisenfeld, A. Bilmes, A. Megrant, R. Barends, J. Kelly, P. Klimov, G. Weiss, J. M. Martinis, and A. V. Ustinov, "Electric field spectroscopy of material defects in transmon qubits", npj Quantum Inf. **5**, 1–6 (2019) (cit. on pp. 22, 39, 152).

[66] M. Neeley, M. Ansmann, R. C. Bialczak, M. Hofheinz, N. Katz, E. Lucero, A. O'Connell, H. Wang, A. N. Cleland, and J. M. Martinis, "Process tomography of quantum memory in a Josephson-phase qubit coupled to a two-level state", Nat. Phys. **4**, 523–526 (2008) (cit. on pp. 22, 39).

[67] Z. L. Wang, Y. P. Zhong, L. J. He, H. Wang, J. M. Martinis, A. N. Cleland, and Q. W. Xie, "Quantum state characterization of a fast tunable superconducting resonator", Appl. Phys. Lett. **102**, 163503 (2013) (cit. on p. 22).

[68] N. Kirsh, E. Svetitsky, A. L. Burin, M. Schechter, and N. Katz, "Revealing the nonlinear response of a tunneling two-level system ensemble using coupled modes", Phys. Rev. Mater. **1**, 012601 (2017) (cit. on p. 22).

[69] G. Andersson, A. L. O. Bilobran, M. Scigliuzzo, M. M. de Lima, J. H. Cole, and P. Delsing, "Acoustic spectral hole-burning in a two-level system ensemble", npj Quantum Inf. **7**, 1–5 (2021) (cit. on p. 22).

[70] T. Capelle, E. Flurin, E. Ivanov, J. Palomo, M. Rosticher, et al., "Probing a Two-Level System Bath via the Frequency Shift of an Off-Resonantly Driven Cavity", Phys. Rev. Appl. **13**, 034022 (2020) (cit. on p. 22).

[71] I. M. Pop, K. Geerlings, G. Catelani, R. J. Schoelkopf, L. I. Glazman, and M. H. Devoret, "Coherent suppression of electromagnetic dissipation due to superconducting quasiparticles", Nature **508**, 369–372 (2014) (cit. on pp. 22, 23, 89, 126, 152).

[72] D. A. Broadway, J.-P. Tetienne, A. Stacey, J. D. A. Wood, D. A. Simpson, L. T. Hall, and L. C. L. Hollenberg, "Quantum probe hyperpolarisation of molecular nuclear spins", Nat. Commun. **9**, 1–8 (2018) (cit. on pp. 22, 25, 45).

[73] H. Bluhm, S. Foletti, D. Mahalu, V. Umansky, and A. Yacoby, "Enhancing the coherence of a spin qubit by operating it as a feedback loop that controls its nuclear spin bath", Phys. Rev. Lett. **105**, 216803 (2010), (cit. on p. 22).

[74] A. P. Vepsäläinen, A. H. Karamlou, J. L. Orrell, A. S. Dogra, B. Loer, et al., "Impact of ionizing radiation on superconducting qubit coherence", Nature **584**, 551–556 (2020) (cit. on pp. 22, 23).

[75] L. Cardani, F. Valenti, N. Casali, G. Catelani, T. Charpentier, et al., "Reducing the impact of radioactivity on quantum circuits in a deep-underground facility", Nat. Commun. **12**, 1–6 (2021) (cit. on p. 22).

[76] S. Schlör, J. Lisenfeld, C. Müller, A. Bilmes, A. Schneider, D. P. Pappas, A. V. Ustinov, and M. Weides, "Correlating Decoherence in Transmon Qubits: Low Frequency Noise by Single Fluctuators", Phys. Rev. Lett. **123**, 190502 (2019) (cit. on p. 22).

[77] M. A. Yurtalan, J. Shi, G. J. K. Flatt, and A. Lupascu, "Characterization of Multilevel Dynamics and Decoherence in a High-Anharmonicity Capacitively Shunted Flux Circuit", Phys. Rev. Appl. **16**, 054051 (2021) (cit. on p. 23).

[78] S. Gustavsson, F. Yan, G. Catelani, J. Bylander, A. Kamal, et al., "Suppressing relaxation in superconducting qubits by quasiparticle pumping", Science **354**, 1573–1577 (2016) (cit. on pp. 23, 24, 93).

[79] H. T. Quan, Y.-x. Liu, C. P. Sun, and F. Nori, "Quantum thermodynamic cycles and quantum heat engines", Phys. Rev. E **76**, 031105 (2007) (cit. on p. 25).

[80] F. Pedregosa, G. Varoquaux, A. Gramfort, V. Michel, B. Thirion, et al., "Scikit-learn: machine learning in Python", Journal of Machine Learning Research **12**, 2825–2830 (2011) (cit. on p. 32).

[81] A. Yang, M. Steger, T. Sekiguchi, M. L. W. Thewalt, T. D. Ladd, K. M. Itoh, H. Riemann, N. V. Abrosimov, P. Becker, and H.-J. Pohl, "Simultaneous subsecond hyperpolarization of the nuclear and electron spins of phosphorus in silicon by optical pumping of exciton transitions", Phys. Rev. Lett. **102**, 257401 (2009) (cit. on p. 36).

[82] H.-J. Wang, C. S. Shin, C. E. Avalos, S. J. Seltzer, D. Budker, A. Pines, and V. S. Bajaj, "Sensitive magnetic control of ensemble nuclear spin hyperpolarization in diamond", Nat. Commun. **4**, 1–7 (2013) (cit. on p. 36).

[83] I. Solomon, "Relaxation Processes in a System of Two Spins", Phys. Rev. **99**, 559–565 (1955) (cit. on pp. 36, 37, 45, 57, 62, 65, 66).

[84] R. Barends, J. Kelly, A. Megrant, D. Sank, E. Jeffrey, et al., "Coherent josephson qubit suitable for scalable quantum integrated circuits", Phys. Rev. Lett. **111**, 080502 (2013) (cit. on pp. 36, 57, 63, 65).

[85] B. Vögeli, "The nuclear Overhauser effect from a quantitative perspective", Prog. Nucl. Magn. Reson. Spectrosc. **78**, 1–46 (2014) (cit. on pp. 37, 45, 46).

[86] M. Spiecker, A. I. Pavlov, A. Shnirman, and I. M. Pop, "Solomon equations for qubit and two-level systems: Insights into non-Poissonian quantum jumps", Phys. Rev. A **109**, 052218 (2024) (cit. on pp. 38, 45, 53, 57, 60, 64, 70, 75, 76, 78, 82–84, 86, 87, 137).

[87] J. Lisenfeld, C. Müller, J. H. Cole, P. Bushev, A. Lukashenko, A. Shnirman, and A. V. Ustinov, "Measuring the Temperature Dependence of Individual Two-Level Systems by Direct Coherent Control", Phys. Rev. Lett. **105**, 230504 (2010) (cit. on pp. 39, 68).

[88] K. H. Lee, G. E. Holmberg, and J. H. Crawford, "Optical and ESR studies of hole centers in γ-irradiated Al2O3", Phys. Status Solidi A **39**, 669–674 (1977) (cit. on p. 39).

[89] S. E. De Graaf, L. Faoro, L. B. Ioffe, S. Mahashabde, J. J. Burnett, T. Lindström, S. E. Kubatkin, A. V. Danilov, and A. Y. Tzalenchuk, "Two-level systems in superconducting quantum devices due to trapped quasiparticles", Sci. Adv. **6**, eabc5055 (2020) (cit. on p. 39).

[90] I. Takmakov, Minimizing the discrimination time for quantum states of an artificial atom, PhD thesis (Karlsruher Institut für Technologie, 2022), DOI: `10.5445/IR/1000142961`, (cit. on pp. 42, 123).

[91] N. Bloembergen, E. M. Purcell, and R. V. Pound, "Relaxation Effects in Nuclear Magnetic Resonance Absorption", Phys. Rev. **73**, 679–712 (1948) (cit. on p. 45).

[92] G. de Lange, Z. H. Wang, D. Ristè, V. V. Dobrovitski, and R. Hanson, "Universal Dynamical Decoupling of a Single Solid-State Spin from a Spin Bath", Science **330**, 60–63 (2010) (cit. on p. 45).

[93] X. Zhu, S. Saito, A. Kemp, K. Kakuyanagi, S.-i. Karimoto, et al., "Coherent coupling of a superconducting flux qubit to an electron spin ensemble in diamond", Nature **478**, 221–224 (2011) (cit. on p. 45).

[94] Y. Kubo, C. Grezes, A. Dewes, T. Umeda, J. Isoya, et al., "Hybrid Quantum Circuit with a Superconducting Qubit Coupled to a Spin Ensemble", Phys. Rev. Lett. **107**, 220501 (2011) (cit. on p. 45).

[95] E. A. Chekhovich, M. N. Makhonin, A. I. Tartakovskii, A. Yacoby, H. Bluhm, K. C. Nowack, and L. M. K. Vandersypen, "Nuclear spin effects in semiconductor quantum dots", Nat. Mater. **12**, 494–504 (2013) (cit. on p. 45).

[96] M. H. Abobeih, J. Cramer, M. A. Bakker, N. Kalb, M. Markham, D. J. Twitchen, and T. H. Taminiau, "One-second coherence for a single electron spin coupled to a multi-qubit nuclear-spin environment", Nat. Commun. **9**, 1–8 (2018) (cit. on p. 45).

[97] M. H. Abobeih, J. Randall, C. E. Bradley, H. P. Bartling, M. A. Bakker, M. J. Degen, M. Markham, D. J. Twitchen, and T. H. Taminiau, "Atomic-scale imaging of a 27-nuclear-spin cluster using a quantum sensor", Nature **576**, 411–415 (2019) (cit. on p. 45).

[98] D. B. R. Dasari, S. Yang, A. Chakrabarti, A. Finkler, G. Kurizki, and J. Wrachtrup, "Anti-Zeno purification of spin baths by quantum probe measurements", Nat. Commun. **13**, 1–10 (2022) (cit. on pp. 45, 79).

[99] J. M. Taylor, A. Imamoglu, and M. D. Lukin, "Controlling a Mesoscopic Spin Environment by Quantum Bit Manipulation", Phys. Rev. Lett. **91**, 246802 (2003) (cit. on p. 45).

[100] E. Barnes, Ł. Cywiński, and S. Das Sarma, "Nonperturbative Master Equation Solution of Central Spin Dephasing Dynamics", Phys. Rev. Lett. **109**, 140403 (2012) (cit. on p. 45).

[101] J. Jing and L.-A. Wu, "Decoherence and control of a qubit in spin baths: an exact master equation study", Sci. Rep. **8**, 1–10 (2018) (cit. on p. 45).

[102] W. H. Zurek, "Environment-induced superselection rules", Phys. Rev. D **26**, 1862–1880 (1982), (cit. on p. 45).

[103] R. Blume-Kohout and W. H. Zurek, "A Simple Example of "Quantum Darwinism": Redundant Information Storage in Many-Spin Environments", Found. Phys. **35**, 1857–1876 (2005) (cit. on p. 45).

[104] B. A. Borgias, M. Gochin, D. J. Kerwood, and T. L. James, "Relaxation matrix analysis of 2D NMR data", Prog. Nucl. Magn. Reson. Spectrosc. **22**, 83–100 (1990) (cit. on p. 45).

[105] H. Friebolin, Basic one- and two-dimensional NMR spectroscopy (Wiley, 2010) (cit. on pp. 45, 67).

[106] A. Kumar, R. Christy Rani Grace, and P. K. Madhu, "Cross-correlations in NMR", Prog. Nucl. Magn. Reson. Spectrosc. **37**, 191–319 (2000) (cit. on pp. 45, 46, 62, 63).

[107] L. G. Werbelow, D. M. Grant, E. P. Black, and J. M. Courtieu, "Magnetization modes and evolution matrices for some simple spin systems in anisotropic media", J. Chem. Phys. **69**, 2407–2419 (1978) (cit. on p. 46).

[108] G. Lindblad, "On the generators of quantum dynamical semigroups", Commun. Math. Phys. **48**, 119–130 (1976) (cit. on p. 46).

[109] V. Gorini, A. Kossakowski, and E. C. G. Sudarshan, "Completely positive dynamical semigroups of N-level systems", J. Math. Phys. **17**, 821–825 (1976) (cit. on p. 46).

[110] N. Staudenmaier, A. Vijayakumar-Sreeja, S. Oviedo-Casado, G. Genov, D. Cohen, et al., "Power-law scaling of correlations in statistically polarised nano-NMR", npj Quantum Inf. **8**, 1–9 (2022) (cit. on p. 46).

[111] P. Bethke, R. P. G. McNeil, J. Ritzmann, T. Botzem, A. Ludwig, A. D. Wieck, and H. Bluhm, "Measurement of Backaction from Electron Spins in a Gate-Defined GaAs Double Quantum dot Coupled to a Mesoscopic Nuclear Spin Bath", Phys. Rev. Lett. **125**, 047701 (2020) (cit. on p. 47).

[112] C. Berke, E. Varvelis, S. Trebst, A. Altland, and D. P. DiVincenzo, "Transmon platform for quantum computing challenged by chaotic fluctuations", Nat. Commun. **13**, 1–10 (2022) (cit. on p. 47).

[113] N. V. Prokof'ev and P. C. E. Stamp, "Theory of the spin bath", Rep. Prog. Phys. **63**, 669 (2000) (cit. on p. 47).

[114] C.-Y. Hsieh and J. Cao, "A unified stochastic formulation of dissipative quantum dynamics. II. Beyond linear response of spin baths", J. Chem. Phys. **148**, 014103 (2018) (cit. on p. 47).

[115] T. Villazon, A. Chandran, and P. W. Claeys, "Integrability and dark states in an anisotropic central spin model", Phys. Rev. Res. **2**, 032052 (2020) (cit. on p. 47).

[116] E. M. Kessler, S. Yelin, M. D. Lukin, J. I. Cirac, and G. Giedke, "Optical Superradiance from Nuclear Spin Environment of Single-Photon Emitters", Phys. Rev. Lett. **104**, 143601 (2010) (cit. on pp. 47, 67).

[117] L. Shao, R. Zhang, W. Lu, Z. Zhang, and X. Wang, "Quantum phase transition in the XXZ central spin model", Phys. Rev. A **107**, 013714 (2023) (cit. on pp. 47, 67).

[118] S. Dooley, F. McCrossan, D. Harland, M. J. Everitt, and T. P. Spiller, "Collapse and revival and cat states with an N-spin system", Phys. Rev. A **87**, 052323 (2013) (cit. on p. 47).

[119] W.-B. He, S. Chesi, H.-Q. Lin, and X.-W. Guan, "Exact quantum dynamics of XXZ central spin problems", Phys. Rev. B **99**, 174308 (2019) (cit. on p. 47).

[120] J. Fan and S. Pang, "Collapse and revival structure of information backflow for a central spin coupled to a finite spin bath", Phys. Rev. A **107**, 022209 (2023) (cit. on p. 47).

[121] Y. Tserkovnyak, "Exceptional points in dissipatively coupled spin dynamics", Phys. Rev. Res. **2**, 013031 (2020) (cit. on p. 47).

[122] H.-P. Breuer and F. Petruccione, The theory of open quantum systems (Oxford University Press, USA, 2002) (cit. on pp. 47, 58, 79, 85).

[123] P. W. Anderson and P. R. Weiss, "Exchange Narrowing in Paramagnetic Resonance", Rev. Mod. Phys. **25**, 269–276 (1953) (cit. on pp. 48, 51).

[124] R. Kubo and K. Tomita, "A General Theory of Magnetic Resonance Absorption", J. Phys. Soc. Jpn. **9**, 888–919 (1954) (cit. on pp. 48, 51).

[125] R. Kubo, "Stochastic Liouville Equations", J. Math. Phys. **4**, 174–183 (1963) (cit. on pp. 51–53).

[126] N. A. Sergeev and M. Olszewski, "Kubo–Anderson oscillator and NMR of solid state", Solid State Nucl. Magn. Reson. **34**, 167–174 (2008) (cit. on pp. 51, 52).

[127] I. Goychuk and P. Hänggi, "Quantum dynamics in strong fluctuating fields", Advances in Physics **54**, 525–584 (2005) (cit. on pp. 52, 53, 57).

[128] M. Goldman, "Formal Theory of Spin–Lattice Relaxation", J. Magn. Reson. **149**, 160–187 (2001) (cit. on pp. 53, 61).

[129] G. Ithier, E. Collin, P. Joyez, P. J. Meeson, D. Vion, et al., "Decoherence in a superconducting quantum bit circuit", Phys. Rev. B **72**, 134519 (2005) (cit. on p. 53).

[130] F. Bloch, "Nuclear Induction", Phys. Rev. **70**, 460–474 (1946) (cit. on p. 53).

[131] M. Am-Shallem, R. Kosloff, and N. Moiseyev, "Exceptional points for parameter estimation in open quantum systems: analysis of the Bloch equations", New J. Phys. **17**, 113036 (2015) (cit. on p. 53).

[132] R. K. Wangsness and F. Bloch, "The Dynamical Theory of Nuclear Induction", Phys. Rev. **89**, 728–739 (1953) (cit. on p. 57).

[133] F. Bloch, "Dynamical Theory of Nuclear Induction. II", Phys. Rev. **102**, 104–135 (1956) (cit. on p. 57).

[134] A. G. Redfield, "On the Theory of Relaxation Processes", IBM J. Res. Dev. **1**, 19–31 (1957) (cit. on pp. 57, 58, 80).

[135] G. Schaller and T. Brandes, "Preservation of positivity by dynamical coarse graining", Phys. Rev. A **78**, 022106 (2008) (cit. on p. 58).

[136] H. J. Carmichael, Statistical Methods in Quantum Optics 1 (Springer, Berlin, Germany, 1999) (cit. on pp. 59, 61, 88).

[137] G. Schaller, Open Quantum Systems Far from Equilibrium (Springer International Publishing, Cham, Switzerland, 2014), (cit. on pp. 59, 163).

[138] U. Weiss, Quantum Dissipative Systems, Second (World Scientific Publishing Company, Singapore, 1999), p. 51 (cit. on p. 60).

[139] G. S. Agarwal, "Rotating-Wave Approximation and Spontaneous Emission", Phys. Rev. A **7**, 1195–1197 (1973) (cit. on p. 61).

[140] M. Esposito and P. Gaspard, "Quantum master equation for a system influencing its environment", Phys. Rev. E **68**, 066112 (2003) (cit. on p. 61).

[141] J. P. Pekola, B. Karimi, M. Cattaneo, and S. Maniscalco, "Long-Time Relaxation of a Finite Spin Bath Linearly Coupled to a Qubit", Open Syst. Inf. Dyn. **30**, 2350009 (2023) (cit. on p. 61).

[142] E. M. Kessler, "Generalized Schrieffer-Wolff formalism for dissipative systems", Phys. Rev. A **86**, 012126 (2012) (cit. on p. 62).

[143] C. Timm, "Time-convolutionless master equation for quantum dots: Perturbative expansion to arbitrary order", Phys. Rev. B **83**, 115416 (2011) (cit. on p. 62).

[144] S. Nakajima, "On Quantum Theory of Transport Phenomena: Steady Diffusion", Prog. Theor. Phys. **20**, 948–959 (1958) (cit. on pp. 62, 64).

[145] R. R. Ernst, G. Bodenhausen, and A. Wokaun, Principles of Nuclear Magnetic Resonance in One and Two Dimensions (Clarendon Press, 1987), p. 54 (cit. on p. 63).

[146] N. Müller, G. Bodenhausen, and R. R. Ernst, "Relaxation-induced violations of coherence transfer selection rules in nuclear magnetic resonance", Journal of Magnetic Resonance (1969) **75**, 297–334 (1987) (cit. on pp. 63, 66).

[147] J. Kowalewski and L. Mäler, Nuclear Spin Relaxation in Liquids: Theory, Experiments, and Applications, Second Edition (Taylor & Francis, Andover, England, UK, 2017), pp. 75–95, (cit. on p. 63).

[148] R. Zwanzig, "Ensemble Method in the Theory of Irreversibility", J. Chem. Phys. **33**, 1338–1341 (1960) (cit. on p. 64).

[149] H. A. M. Leymann, A. Foerster, and J. Wiersig, "Expectation value based equation-of-motion approach for open quantum systems: A general formalism", Phys. Rev. B **89**, 085308 (2014) (cit. on p. 66).

[150] R. Röhrig, P. Schering, L. B. Gravert, B. Fauseweh, and G. S. Uhrig, "Quantum mechanical treatment of large spin baths", Phys. Rev. B **97**, 165431 (2018) (cit. on p. 66).

[151] M. Lindberg, Y. Z. Hu, R. Binder, and S. W. Koch, "$\chi^{(3)}$ formalism in optically excited semiconductors and its applications in four-wave-mixing spectroscopy", Phys. Rev. B **50**, 18060–18072 (1994) (cit. on p. 66).

[152] P. Gartner, "Two-level laser: Analytical results and the laser transition", Phys. Rev. A **84**, 053804 (2011) (cit. on p. 66).

[153] B. Fauseweh, P. Schering, J. Hüdepohl, and G. S. Uhrig, "Efficient algorithms for the dynamics of large and infinite classical central spin models", Phys. Rev. B **96**, 054415 (2017) (cit. on p. 66).

[154] C. Müller, A. Shnirman, and Y. Makhlin, "Relaxation of Josephson qubits due to strong coupling to two-level systems", Phys. Rev. B **80**, 134517 (2009) (cit. on p. 67).

[155] Y. Shalibo, Y. Rofe, D. Shwa, F. Zeides, M. Neeley, J. M. Martinis, and N. Katz, "Lifetime and Coherence of Two-Level Defects in a Josephson Junction", Phys. Rev. Lett. **105**, 177001 (2010) (cit. on p. 68).

[156] H. Zhang, S. Chakram, T. Roy, N. Earnest, Y. Lu, Z. Huang, D. K. Weiss, J. Koch, and D. I. Schuster, "Universal Fast-Flux Control of a Coherent, Low-Frequency Qubit", Phys. Rev. X **11**, 011010 (2021) (cit. on pp. 68, 157).

[157] L. Shen and B. W. Suter, "Bounds for Eigenvalues of Arrowhead Matrices and Their Applications to Hub Matrices and Wireless Communications", EURASIP J. Adv. Signal Process. **2009**, 1–12 (2009) (cit. on p. 69).

[158] N. Jakovčević Stor, I. Slapničar, and J. L. Barlow, "Accurate eigenvalue decomposition of real symmetric arrowhead matrices and applications", Linear Algebra Appl. **464**, 62–89 (2015) (cit. on pp. 69, 71).

[159] T. Thorbeck, Z. Xiao, A. Kamal, and L. C. G. Govia, "Readout-Induced Suppression and Enhancement of Superconducting Qubit Lifetimes", Phys. Rev. Lett. **132**, 090602 (2024) (cit. on p. 77).

[160] M. Spiecker, A. Pavlov, A. Shnirman, and I. M. Pop, Solomon equations for qubit and two-level systems: Insights into non-Poissonian quantum jumps [Dataset], (2023), DOI: `10.5281/zenodo.8114088` (cit. on p. 78).

[161] H. M. Wiseman and L. Diósi, "Complete parameterization, and invariance, of diffusive quantum trajectories for Markovian open systems", Chem. Phys. **268**, 91–104 (2001) (cit. on pp. 79, 84, 85).

[162] H. M. Wiseman and G. J. Milburn, "Interpretation of quantum jump and diffusion processes illustrated on the Bloch sphere", Phys. Rev. A **47**, 1652–1666 (1993) (cit. on pp. 79, 85).

[163] K. Mølmer, Y. Castin, and J. Dalibard, "Monte Carlo wave-function method in quantum optics", J. Opt. Soc. Am. B, JOSAB **10**, 524–538 (1993) (cit. on p. 84).

[164] N. Gisin and I. C. Percival, "The quantum-state diffusion model applied to open systems", J. Phys. A: Math. Gen. **25**, 5677 (1992) (cit. on p. 84).

[165] Y. Castin, J. Dalibard, and K. Mølmer, "A wave function approach to dissipative processes", AIP Conf. Proc. **275**, 143–156 (1993) (cit. on p. 85).

[166] J. C. Bergquist, R. G. Hulet, W. M. Itano, and D. J. Wineland, "Observation of Quantum Jumps in a Single Atom", Phys. Rev. Lett. **57**, 1699–1702 (1986) (cit. on pp. 88, 137).

[167] M. B. Plenio and P. L. Knight, "The quantum-jump approach to dissipative dynamics in quantum optics", Rev. Mod. Phys. **70**, 101–144 (1998) (cit. on p. 88).

[168] H. Carmichael, "A tale of quantum jumps", nzsr **72**, 31–34 (2015) (cit. on p. 88).

[169] T. M. Hazard, A. Gyenis, A. Di Paolo, A. T. Asfaw, S. A. Lyon, A. Blais, and A. A. Houck, "Nanowire Superinductance Fluxonium Qubit", Phys. Rev. Lett. **122**, 010504 (2019) (cit. on p. 93).

[170] R. Gao, H.-S. Ku, H. Deng, W. Yu, T. Xia, et al., "Ultrahigh Kinetic Inductance Superconducting Materials from Spinodal Decomposition", Adv. Mater. **34**, 2201268 (2022) (cit. on p. 93).

[171] D. Rieger, S. Günzler, M. Spiecker, A. Nambisan, W. Wernsdorfer, and I. M. Pop, "Fano Interference in Microwave Resonator Measurements", Phys. Rev. Appl. **20**, 014059 (2023) (cit. on p. 93).

[172] A. Jayaraman, A. V. Danilov, J. Bylander, and S. E. Kubatkin, "Loss and decoherence in superconducting circuits on silicon: Insights from electron spin resonance", Phys. Rev. Appl. **22**, 014030 (2024) (cit. on p. 95).

[173] M. Odeh, K. Godeneli, E. Li, R. Tangirala, H. Zhou, X. Zhang, Z.-H. Zhang, and A. Sipahigil, "Non-Markovian dynamics of a superconducting qubit in a phononic bandgap", (2023), DOI: 10.48550/arXiv.2312.01031 (cit. on p. 95).

[174] B. Yurke and J. S. Denker, "Quantum network theory", Phys. Rev. A **29**, 1419–1437 (1984) (cit. on p. 99).

[175] M. H. Devoret, Quantum Fluctuations, Les Houches, Session LXIII, 1995 (Elsevier, Amsterdam, 1997), pp. 133–135 (cit. on p. 99).

[176] U. Vool and M. Devoret, "Introduction to quantum electromagnetic circuits", Int. J. Circuit Theory Appl. **45**, 897–934 (2017) (cit. on pp. 99, 102).

[177] S. E. Nigg, H. Paik, B. Vlastakis, G. Kirchmair, S. Shankar, L. Frunzio, M. H. Devoret, R. J. Schoelkopf, and S. M. Girvin, "Black-Box Superconducting Circuit Quantization", Phys. Rev. Lett. **108**, 240502 (2012) (cit. on p. 99).

[178] D. Ding, H.-S. Ku, Y. Shi, and H.-H. Zhao, "Free-mode removal and mode decoupling for simulating general superconducting quantum circuits", Phys. Rev. B **103**, 174501 (2021) (cit. on p. 99).

[179] Z. K. Minev, Z. Leghtas, S. O. Mundhada, L. Christakis, I. M. Pop, and M. H. Devoret, "Energy-participation quantization of Josephson circuits", npj Quantum Inf. **7**, 1–11 (2021) (cit. on p. 99).

[180] R.-P. Riwar and D. P. DiVincenzo, "Circuit quantization with time-dependent magnetic fields for realistic geometries", npj Quantum Inf. **8**, 1–12 (2022) (cit. on p. 99).

[181] I. L. Egusquiza and A. Parra-Rodriguez, "Algebraic canonical quantization of lumped superconducting networks", Phys. Rev. B **106**, 024510 (2022) (cit. on pp. 99, 100).

[182] M. Rymarz and D. P. DiVincenzo, "Consistent Quantization of Nearly Singular Superconducting Circuits", Phys. Rev. X **13**, 021017 (2023) (cit. on pp. 99, 100, 109).

[183] A. Osborne, T. Larson, S. G. Jones, R. W. Simmonds, A. Gyenis, and A. Lucas, "Symplectic Geometry and Circuit Quantization", PRX Quantum **5**, 020309 (2024) (cit. on p. 99).

[184] T. Rajabzadeh, Z. Wang, N. Lee, T. Makihara, Y. Guo, and A. H. Safavi-Naeini, "Analysis of arbitrary superconducting quantum circuits accompanied by a Python package: SQcircuit", Quantum **7**, 1118 (2023) (cit. on p. 99).

[185] S. P. Chitta, T. Zhao, Z. Huang, I. Mondragon-Shem, and J. Koch, "Computer-aided quantization and numerical analysis of superconducting circuits", New J. Phys. **24**, 103020 (2022) (cit. on p. 99).

[186] J. Bryon, D. K. Weiss, X. You, S. Sussman, X. Croot, Z. Huang, J. Koch, and A. A. Houck, "Time-Dependent Magnetic Flux in Devices for Circuit Quantum Electrodynamics", Phys. Rev. Appl. **19**, 034031 (2023) (cit. on p. 99).

[187] H. J. Groenewold and H. J. Groenewold, On the principles of elementary quantum mechanics (Springer, 1946) (cit. on p. 99).

[188] U. Eckern, G. Schön, and V. Ambegaokar, "Quantum dynamics of a superconducting tunnel junction", Phys. Rev. B **30**, 6419–6431 (1984) (cit. on pp. 99, 100, 123).

[189] W. C. Smith, M. Villiers, A. Marquet, J. Palomo, M. R. Delbecq, T. Kontos, P. Campagne-Ibarcq, B. Douçot, and Z. Leghtas, "Magnifying Quantum Phase Fluctuations with Cooper-Pair Pairing", Phys. Rev. X **12**, 021002 (2022) (cit. on pp. 100, 125).

[190] A. Wallraff, A. Lukashenko, J. Lisenfeld, A. Kemp, M. V. Fistul, Y. Koval, and A. V. Ustinov, "Quantum dynamics of a single vortex", Nature **425**, 155–158 (2003) (cit. on p. 100).

[191] M. M. Wildermuth, Quantum Tunneling of Josephson Vortices in High-Impedance Long Junctions, PhD thesis (Karlsruher Institut für Technologie, 2023), DOI: 10.5445/IR/1000158804 (cit. on p. 100).

[192] E. P. Gross, "Structure of a quantized vortex in boson systems", Nuovo Cim. **20**, 454–477 (1961) (cit. on p. 100).

[193] L. P. Pitaevskii, "Vortex lines in an imperfect bose gas", Sov. Phys. JETP **13**, 451–454 (1961) (cit. on p. 100).

[194] R. Lopes, C. Eigen, N. Navon, D. Clément, R. P. Smith, and Z. Hadzibabic, "Quantum Depletion of a Homogeneous Bose-Einstein Condensate", Phys. Rev. Lett. **119**, 190404 (2017) (cit. on p. 100).

[195] M. Tinkham, Introduction to Superconductivity (Courier Corporation, 2004) (cit. on pp. 101, 103).

[196] S. Alwyn, Encyclopedia of Nonlinear Science (Routledge, New York, USA, 2015) (cit. on p. 101).

[197] M. Peruzzo, A. Trioni, F. Hassani, M. Zemlicka, and J. M. Fink, "Surpassing the Resistance Quantum with a Geometric Superinductor", Phys. Rev. Appl. **14**, 044055 (2020) (cit. on p. 102).

[198] A. Glezer Moshe, G. Tuvia, S. Avraham, E. Farber, and G. Deutscher, "Tunneling study in granular aluminum near the Mott metal-to-insulator transition", Phys. Rev. B **104**, 054508 (2021) (cit. on pp. 103, 119).

[199] D. Willsch, D. Rieger, P. Winkel, M. Willsch, C. Dickel, et al., "Observation of Josephson harmonics in tunnel junctions", Nat. Phys. 1–7 (2024) (cit. on pp. 103, 107).

[200] B. D. Josephson, "Coupled Superconductors", Rev. Mod. Phys. **36**, 216–220 (1964) (cit. on p. 103).

[201] A. A. Golubov, M. Y. Kupriyanov, and E. Il'ichev, "The current-phase relation in Josephson junctions", Rev. Mod. Phys. **76**, 411–469 (2004) (cit. on p. 103).

[202] N. A. Masluk, I. M. Pop, A. Kamal, Z. K. Minev, and M. H. Devoret, "Microwave Characterization of Josephson Junction Arrays: Implementing a Low Loss Superinductance", Phys. Rev. Lett. **109**, 137002 (2012) (cit. on pp. 103, 113).

[203] D. Gusenkova, F. Valenti, M. Spiecker, S. Günzler, P. Paluch, et al., "Operating in a deep underground facility improves the locking of gradiometric fluxonium qubits at the sweet spots", Appl. Phys. Lett. **120**, 054001 (2022) (cit. on pp. 103, 115, 120, 122, 124).

[204] M. G. Castellano, F. Chiarello, G. Torrioli, and P. Carelli, "Static flux bias of a flux qubit using persistent current trapping", Supercond. Sci. Technol. **19**, 1158 (2006) (cit. on p. 103).

[205] L. Lorenz, "Ueber die Fortpflanzung der Electricität", Ann. Phys. **243**, 161–193 (1879) (cit. on p. 104).

[206] H. Nagaoka, "The Inductance Coefficients of Solenoids", Journal of the College of Science, Imperial University of Tokyo, Japan **27**, 1–33 (1909) (cit. on p. 104).

[207] M. D. Croitoru, S. V. Mironov, B. Lounis, and A. I. Buzdin, "Toward the Light-Operated Superconducting Devices: Circularly Polarized Radiation Manipulates the Current-Carrying States in Superconducting Rings", Adv. Quantum Technol. **5**, 2200054 (2022) (cit. on p. 105).

[208] J. Koch, T. M. Yu, J. Gambetta, A. A. Houck, D. I. Schuster, J. Majer, A. Blais, M. H. Devoret, S. M. Girvin, and R. J. Schoelkopf, "Charge-insensitive qubit design derived from the Cooper pair box", Phys. Rev. A **76**, 042319 (2007) (cit. on pp. 105, 107, 108).

[209] T. A. Fulton and G. J. Dolan, "Observation of single-electron charging effects in small tunnel junctions", Phys. Rev. Lett. **59**, 109–112 (1987) (cit. on p. 105).

[210] A. Schneider, T. Wolz, M. Pfirrmann, M. Spiecker, H. Rotzinger, A. V. Ustinov, and M. Weides, "Transmon qubit in a magnetic field: Evolution of coherence and transition frequency", Phys. Rev. Res. **1**, 023003 (2019) (cit. on p. 109).

[211] I. Chiorescu, Y. Nakamura, C. J. P. M. Harmans, and J. E. Mooij, "Coherent Quantum Dynamics of a Superconducting Flux Qubit", Science **299**, 1869–1871 (2003) (cit. on p. 113).

[212] N. Maleeva, L. Grünhaupt, T. Klein, F. Levy-Bertrand, O. Dupre, et al., "Circuit quantum electrodynamics of granular aluminum resonators", Nat. Commun. **9**, 1–7 (2018) (cit. on pp. 113, 117–119, 123).

[213] J. Stevens, Effect of the Environment on Fluxonium Qubits and Thermodynamics of Quantum Measurement, PhD thesis (Université de Lyon, 2021), (cit. on p. 115).

[214] U. S. Pracht, N. Bachar, L. Benfatto, G. Deutscher, E. Farber, M. Dressel, and M. Scheffler, "Enhanced Cooper pairing versus suppressed phase coherence shaping the superconducting dome in coupled aluminum nanograins", Phys. Rev. B **93**, 100503 (2016) (cit. on p. 119).

[215] N. E. Frattini, U. Vool, S. Shankar, A. Narla, K. M. Sliwa, and M. H. Devoret, "3-wave mixing Josephson dipole element", Appl. Phys. Lett. **110**, 222603 (2017) (cit. on p. 120).

[216] Y. Srivastava and A. Widom, "Quantum electrodynamic processes in electrical engineering circuits", Phys. Rep. **148**, 1–65 (1987) (cit. on p. 123).

[217] M. J. Schwarz, J. Goetz, Z. Jiang, T. Niemczyk, F. Deppe, A. Marx, and R. Gross, "Gradiometric flux qubits with a tunable gap", New J. Phys. **15**, 045001 (2013) (cit. on p. 124).

[218] D. Gusenkova, Quantum non-Demolition Readout of Superconducting Artificial Atoms, PhD thesis (Karlsruher Institut für Technologie, 2021), DOI: 10.5445/IR/1000140969 (cit. on p. 124).

[219] W. C. Smith, A. Kou, U. Vool, I. M. Pop, L. Frunzio, R. J. Schoelkopf, and M. H. Devoret, "Quantization of inductively shunted superconducting circuits", Phys. Rev. B **94**, 144507 (2016) (cit. on pp. 127, 141).

[220] I. S. Gradshteyn and I. M. Ryzhik, Table of integrals, series, and products (Academic press, 2014) (cit. on p. 128).

[221] N. Earnest, S. Chakram, Y. Lu, N. Irons, R. K. Naik, et al., "Realization of a Λ System with Metastable States of a Capacitively Shunted Fluxonium", Phys. Rev. Lett. **120**, 150504 (2018) (cit. on p. 129).

[222] I. V. Pechenezhskiy, R. A. Mencia, L. B. Nguyen, Y.-H. Lin, and V. E. Manucharyan, "The superconducting quasicharge qubit", Nature **585**, 368–371 (2020) (cit. on pp. 113, 129, 133).

[223] A. Petrescu, H. E. Türeci, A. V. Ustinov, and I. M. Pop, "Fluxon-based quantum simulation in circuit QED", Phys. Rev. B **98**, 174505 (2018) (cit. on p. 129).

[224] D. K. Weiss, W. DeGottardi, J. Koch, and D. G. Ferguson, "Variational tight-binding method for simulating large superconducting circuits", Phys. Rev. Res. **3**, 033244 (2021) (cit. on p. 130).

[225] V. B. Braginsky, Y. I. Vorontsov, and K. S. Thorne, "Quantum Nondemolition Measurements", Science **209**, 547–557 (1980) (cit. on p. 137).

[226] W. Nagourney, J. Sandberg, and H. Dehmelt, "Shelved optical electron amplifier: Observation of quantum jumps", Phys. Rev. Lett. **56**, 2797–2799 (1986) (cit. on p. 137).

[227] T. Sauter, W. Neuhauser, R. Blatt, and P. E. Toschek, "Observation of Quantum Jumps", Phys. Rev. Lett. **57**, 1696–1698 (1986) (cit. on p. 137).

[228] G. Nogues, A. Rauschenbeutel, S. Osnaghi, M. Brune, J. M. Raimond, and S. Haroche, "Seeing a single photon without destroying it", Nature **400**, 239–242 (1999) (cit. on p. 137).

[229] S. Peil and G. Gabrielse, "Observing the Quantum Limit of an Electron Cyclotron: QND Measurements of Quantum Jumps between Fock States", Phys. Rev. Lett. **83**, 1287–1290 (1999) (cit. on p. 137).

[230] C. M. Caves, K. S. Thorne, R. W. P. Drever, V. D. Sandberg, and M. Zimmermann, "On the measurement of a weak classical force coupled to a quantum-mechanical oscillator. I. Issues of principle", Rev. Mod. Phys. **52**, 341–392 (1980) (cit. on p. 137).

[231] S. Roy, J. T. Chalker, I. V. Gornyi, and Y. Gefen, "Measurement-induced steering of quantum systems", Phys. Rev. Res. **2**, 033347 (2020) (cit. on p. 137).

[232] S. M. Carroll and J. Lodman, "Energy Non-conservation in Quantum Mechanics", Found. Phys. **51**, 1–15 (2021) (cit. on p. 137).

[233] Y. Aharonov, S. Popescu, and D. Rohrlich, "Conservation laws and the foundations of quantum mechanics", Proc. Natl. Acad. Sci. U.S.A. **120**, e2220810120 (2023) (cit. on p. 137).

[234] S. Filipp, P. Maurer, P. J. Leek, M. Baur, R. Bianchetti, et al., "Two-Qubit State Tomography Using a Joint Dispersive Readout", Phys. Rev. Lett. **102**, 200402 (2009) (cit. on p. 138).

[235] Q. Xie, H. Zhong, M. T. Batchelor, and C. Lee, "The quantum Rabi model: solution and dynamics", J. Phys. A: Math. Theor. **50**, 113001 (2017) (cit. on p. 138).

[236] I. Takmakov, P. Winkel, F. Foroughi, L. Planat, D. Gusenkova, et al., "Minimizing the Discrimination Time for Quantum States of an Artificial Atom", Phys. Rev. Appl. **15**, 064029 (2021) (cit. on p. 139).

[237] M. Spiecker, Black Forest Quantum Circuits (Version v0.1.1) [Software], (2024), DOI: 10.5281/zenodo.13695982, URL: https://github.com/MartinSpiecker/bfqcircuits (cit. on p. 144).

[238] R. Zhao, S. Park, T. Zhao, M. Bal, C. R. H. McRae, J. Long, and D. P. Pappas, "Merged-Element Transmon", Phys. Rev. Appl. **14**, 064006 (2020) (cit. on p. 152).

[239] H. J. Mamin, E. Huang, S. Carnevale, C. T. Rettner, N. Arellano, et al., "Merged-Element Transmons: Design and Qubit Performance", Phys. Rev. Appl. **16**, 024023 (2021) (cit. on p. 152).

[240] A. Kou, W. C. Smith, U. Vool, I. M. Pop, K. M. Sliwa, M. Hatridge, L. Frunzio, and M. H. Devoret, "Simultaneous Monitoring of Fluxonium Qubits in a Waveguide", Phys. Rev. Appl. **9**, 064022 (2018) (cit. on p. 152).

[241] F. Lecocq, I. M. Pop, Z. Peng, I. Matei, T. Crozes, T. Fournier, C. Naud, W. Guichard, and O. Buisson, "Junction fabrication by shadow evaporation without a suspended bridge", Nanotechnology **22**, 315302 (2011) (cit. on p. 152).

[242] G. J. Dolan, "Offset masks for lift-off photoprocessing", Appl. Phys. Lett. **31**, 337–339 (1977) (cit. on pp. 152, 153).

[243] P. S. Mundada, A. Gyenis, Z. Huang, J. Koch, and A. A. Houck, "Floquet-Engineered Enhancement of Coherence Times in a Driven Fluxonium Qubit", Phys. Rev. Appl. **14**, 054033 (2020) (cit. on p. 157).

[244] L. B. Nguyen, G. Koolstra, Y. Kim, A. Morvan, T. Chistolini, et al., "Blueprint for a High-Performance Fluxonium Quantum Processor", PRX Quantum **3**, 037001 (2022) (cit. on p. 157).

[245] F. Henriques, F. Valenti, T. Charpentier, M. Lagoin, C. Gouriou, et al., "Phonon traps reduce the quasiparticle density in superconducting circuits", Appl. Phys. Lett. **115**, 212601 (2019) (cit. on p. 158).

[246] A. Bargerbos, L. J. Splitthoff, M. Pita-Vidal, J. J. Wesdorp, Y. Liu, P. Krogstrup, L. P. Kouwenhoven, C. K. Andersen, and L. Grünhaupt, "Mitigation of Quasiparticle Loss in Superconducting Qubits by Phonon Scattering", Phys. Rev. Appl. **19**, 024014 (2023) (cit. on p. 158).

[247] V. Iaia, J. Ku, A. Ballard, C. P. Larson, E. Yelton, C. H. Liu, S. Patel, R. McDermott, and B. L. T. Plourde, "Phonon downconversion to suppress correlated errors in superconducting qubits", Nat. Commun. **13**, 1–7 (2022) (cit. on p. 158).

List of Publications

Authors marked with † have contributed equally.

M. Spiecker, A. I. Pavlov, A. Shnirman, and I. M. Pop, "Solomon equations for qubit and two-level systems: Insights into non-Poissonian quantum jumps", Phys. Rev. A **109**, 052218 (2024).

D. Rieger, S. Günzler, **M. Spiecker**, A. Nambisan, W. Wernsdorfer, and I. M. Pop, "Fano Interference in Microwave Resonator Measurements", Phys. Rev. Appl. **20**, 014059 (2023).

M. Spiecker, P. Paluch, N. Gosling, N. Drucker, S. Matityahu, D. Gusenkova, S. Günzler, D. Rieger, I. Takmakov, F. Valenti, P. Winkel, R. Gebauer, O. Sander, G. Catelani, A. Shnirman, A. V. Ustinov, W. Wernsdorfer, Y. Cohen, and I. M. Pop, "Two-level system hyperpolarization using a quantum Szilard engine", Nat. Phys. **19**, 1320–1325 (2023).

D. Rieger, S. Günzler, **M. Spiecker**, P. Paluch, P. Winkel, L. Hahn, J. K. Hohmann, A. Bacher, W. Wernsdorfer, and I. M. Pop, "Granular aluminium nanojunction fluxonium qubit", Nat. Mater. **22**, 194–199 (2023).

D. Gusenkova, F. Valenti, **M. Spiecker**, S. Günzler, P. Paluch, D. Rieger, L.-M. Pioraş-Ţimbolmaş, L. P. Zârbo, N. Casali, I. Colantoni, A. Cruciani, S. Pirro, L. Cardani, A. Petrescu, W. Wernsdorfer, P. Winkel, and I. M. Pop, "Operating in a deep underground facility improves the locking of gradiometric fluxonium qubits at the sweet spots", Appl. Phys. Lett. **120**, 054001 (2022).

S. Günzler, P. Winkel, D. Rieger, K. Borisov, **M. Spiecker**, A. V. Ustinov, I. M. Pop, and W. Wernsdorfer, "Superconducting microwave magnetometer for absolute flux detection", (2021), DOI: `10.48550/arXiv.2107.05929`.

I. Takmakov, P. Winkel, F. Foroughi, L. Planat, D. Gusenkova, **M. Spiecker**, D. Rieger, L. Grünhaupt, A. V. Ustinov, W. Wernsdorfer, I. M. Pop, and N. Roch, "Minimizing the Discrimination Time for Quantum States of an Artificial Atom", Phys. Rev. Appl. **15**, 064029 (2021).

D. Gusenkova†, **M. Spiecker†**, R. Gebauer, M. Willsch, D. Willsch, F. Valenti, N. Karcher, L. Grünhaupt, I. Takmakov, P. Winkel, D. Rieger, A. V. Ustinov, N. Roch, W. Wernsdorfer, K. Michielsen, O. Sander, and I. M. Pop, "Quantum Nondemolition Dispersive Readout of a Superconducting Artificial Atom Using Large Photon Numbers", Phys. Rev. Appl. **15**, 064030 (2021).

K. Borisov, D. Rieger, P. Winkel, F. Henriques, F. Valenti, A. Ionita, M. Wessbecher, **M. Spiecker**, D. Gusenkova, I. M. Pop, and W. Wernsdorfer, "Superconducting granular aluminum resonators resilient to magnetic fields up to 1 Tesla", Appl. Phys. Lett. **117**, 120502 (2020).

P. Winkel, K. Borisov, L. Grünhaupt, D. Rieger, **M. Spiecker**, F. Valenti, A. V. Ustinov, W. Wernsdorfer, and I. M. Pop, "Implementation of a Transmon Qubit Using Superconducting Granular Aluminum", Phys. Rev. X **10**, 031032 (2020).

R. Gebauer, N. Karcher, D. Gusenkova, **M. Spiecker**, L. Grünhaupt, I. Takmakov, P. Winkel, L. Planat, N. Roch, W. Wernsdorfer, A. V. Ustinov, M. Weber, M. Weides, I. M. Pop, and O. Sander, "State preparation of a fluxonium qubit with feedback from a custom FPGA-based platform", AIP Conf. Proc. **2241**, 020015 (2020).

A. Schneider, T. Wolz, M. Pfirrmann, **M. Spiecker**, H. Rotzinger, A. V. Ustinov, and M. Weides, "Transmon qubit in a magnetic field: Evolution of coherence and transition frequency", Phys. Rev. Res. **1**, 023003 (2019).

L. Grünhaupt†, **M. Spiecker†**, D. Gusenkova, N. Maleeva, S. T. Skacel, I. Takmakov, F. Valenti, P. Winkel, H. Rotzinger, W. Wernsdorfer, A. V. Ustinov, and I. M. Pop, "Granular aluminium as a superconducting material for high-impedance quantum circuits", Nat. Mater. **18**, 816–819 (2019).

Publications not related to the dissertation

R. Spiecker, P. Pfeiffer, A. Biswal, M. Shcherbinin, **M. Spiecker**, H. Hessdorfer, M. Hurst, Y. Zharov, V. Bellucci, T. Faragó, M. Zuber, A. Herz, A. Cecilia, M. Czyzycki, C. S. B. Dias, D. Novikov, L. Krogmann, E. Hamann, T. van de Kamp, and T. Baumbach, "Dose-efficient *in vivo* X-ray phase contrast imaging at micrometer resolution by Bragg magnifiers", Optica **10**, 1633–1640 (2023).

R. Spiecker, **M. Spiecker**, A. Biswal, M. Shcherbinin, and T. Baumbach, "The Bragg demagnifier: X-ray imaging with kilometer propagation distance within a meter", (2023), DOI: `10.48550/arXiv.2310.16771`.

R. Pretzsch, M. Dries, S. Hettler, **M. Spiecker**, M. Obermair, and D. Gerthsen, "Investigation of hole-free phase plate performance in transmission electron microscopy under different operation conditions by experiments and simulations", Adv. Struct. Chem. Imag. **5**, 1–11 (2019).

I. Gregor, **M. Spiecker**, R. Petrovsky, J. Großhans, R. Ros, and J. Enderlein, "Rapid nonlinear image scanning microscopy", Nat. Methods **14**, 1087–1089 (2017).

Acknowledgments

After all these intense and exciting years I would like to take the opportunity to thank everyone who accompanied me on this journey and who has contributed to the success of this work in various ways.

First and foremost, I would like to thank Ioan Pop for being my supervisor. I can still remember the moment when you sparked my initial curiosity about this research topic. We had you as a special guest in our lecture, when you proudly presented your self-made artificial atom, which looked more like a fish skeleton to me. Beyond that, it was your enthusiasm and your lighthearted nature that convinced me to join your group. Your joyful and motivating way of managing the group is truly remarkable. I am grateful for everything I could learn from you. Thank you for your support and trust, for all the discussions we had about physics and life in general and for numerous hours you spent polishing papers and conference talks. Moreover, I would like to thank you for sending me to various conferences where I could present and discuss my research.

Next, I express my gratitude to Alexander Shnirman. Thank you for your interest in my work and for serving as my second supervisor. It was always exciting and a pleasure to discuss with you at the whiteboard. Your long-term experience in the field was certainly enriching, both for me personally and for this work.

Furthermore, I am grateful to thank Stefan Filipp for the time and effort he invested in reviewing my thesis and for attending my doctoral defense.

I thank all my colleagues in Ioan's group for a fantastic working atmosphere, for lasting memories of numerous conferences, various hiking seminars, the summer school in Les Houches, as well as for the social activities outside of work. Especially, I would like to thank Patrick Paluch for all the time we spent together in the office and in the lab. I could always rely on your help. Also, I will never forget your legendary parties. Daria Gusenkova, it was great working with you. I enjoyed the calm atmosphere that you created in the office, which was only interrupted by the trenchant humor of Francesco Valenti, whom I would also like to thank for organizing several lovely hiking seminars in the Italian Alps. Thanks to Lukas Grünhaupt for introducing me to the field and showing me how to operate a wet fridge. Thank you Nicolas Gosling for helping me with supplementary measurements. Inspired by your enthusiasm for open quantum systems, I took my first tentative steps into this messy field. It would be impossible to imagine my PhD time without Dennis Rieger. Besides all the nice physics discussions we had, you were always the right person to go to when I was stuck with computer problems. Your remarkable processing speed not only became visible on your screen, but also during our summerly roundnet sessions. Your good vibes were also spread by our institute Christmas brass band. Filled with similar energy and drive, Mitchell Field joined our group at the end of my PhD. I was very lucky to have you by my side. Thank you for proofreading parts of my manuscript and for answering any language questions. I extend my thanks to Patrick Winkel, Ivan Takmakov, and Nicolas Zapata for providing a steady flow of fresh parametric amplifiers that were crucial for this work.

The experiments would not have been possible without the great company of Niv Drucker, who closely followed the experimental developments and could always provide me with fresh software updates and workarounds.

Special thanks go to Andrei Pavlov. Thank you for all the interesting discussions we had and for your help with finalizing and publishing the theoretical part of this work.

I would also like to mention the fruitful environment spread by all the colleagues from the neighboring groups at our institute. In particular, I want to thank Micha Wildermuth, Tim Wolz, and André Schneider for their company and collaboration. I also thank Alexey Ustinov for letting me participate in his ski seminars and Wolfgang Wernsdorfer for letting me use his brand-new fridges.

Furthermore, I appreciate the support of our administration and IT, especially Steffi Baatz, Claudia Alaya, Dorothea Trautmann and Lars Behrens, as well as the supply by the mechanical workshops. Thank you Michael Meyer, Frank Landhäußler and Maik Nothdurft for your precise work. A huge thanks also goes to Hans-Willi Pensl and Christoph Sürgers for taking care of our old helium liquefier. Regarding the nano-fabrication, I owe a big thank you to Silvia Diewald for her dedication in optimizing the e-beam writing process as well as to Lucas Radtke for his support in the cleanroom.

In general, I enjoyed a lot the thriving and constructive atmosphere in our community. Therefore, I want to express a general thank you to all the scientists that I have met on my way. Keep up the good work and spirit!

Finally, I want to thank my family, especially my parents for supporting me throughout my life. I am grateful for everything I could and still can learn from you. To my brothers and sisters-in-law, thank you for all the refreshing distractions from work and for reminding me that there is more to life than physics. Lastly, to my wonderful wife, Rebecca, thank you for your love, your patience, all your help, and for going through this PhD time together. I look forward to our next adventures!

Experimental Condensed Matter Physics
(ISSN 2191-9925)

Band 1 Alexey Feofanov
Experiments on flux qubits with pi-shifters.
ISBN 978-3-86644-644-1

Band 2 Stefan Schmaus
Spintronics with individual metal-organic molecules.
ISBN 978-3-86644-649-6

Band 3 Marc Müller
Elektrischer Leitwert von magnetostriktiven Dy-Nanokontakten.
ISBN 978-3-86644-726-4

Band 4 Torben Peichl
Einfluss mechanischer Deformation auf atomare Tunnelsysteme – untersucht mit Josephson Phasen-Qubits.
ISBN 978-3-86644-837-7

Band 5 Dominik Stöffler
Herstellung dünner metallischer Brücken durch Elektromigration und Charakterisierung mit Rastersondentechniken.
ISBN 978-3-86644-843-8

Band 6 Tihomir Tomanic
Untersuchung des elektronischen Oberflächenzustands von Ag-Inseln auf supraleitendem Niob (110).
ISBN 978-3-86644-898-8

Band 7 Lukas Gerhard
Magnetoelectric coupling at metal surfaces.
ISBN 978-3-7315-0063-6

Band 8 Nicht erschienen.

Band 9 Jochen Zimmer
Cooper pair transport in arrays of Josephson junctions.
ISBN 978-3-7315-0130-5

Band 10 Oliver Berg
Elektrischer Transport durch Nanokontakte von Selten-Erd-Metallen.
ISBN 978-3-7315-0209-8

Band 11 Grigorij Jur'evic Grabovskij
Investigation of coherent microscopic defects inside the tunneling barrier of a Josephson junction.
ISBN 978-3-7315-0210-4

Die Bände sind unter www.ksp.kit.edu als PDF frei verfügbar oder als Druckausgabe bestellbar.

Band 12 Cornelius Thiele
STM Characterization of Phenylene-Ethynylene Oligomers on Au(111) and their Integration into Carbon Nanotube Nanogaps.
ISBN 978-3-7315-0235-7

Band 13 Michael Peter Schackert
Scanning Tunneling Spectroscopy on Electron-Boson Interactions in Superconductors.
ISBN 978-3-7315-0238-8

Band 14 Susanne Butz
One-Dimensional Tunable Josephson Metamaterials.
ISBN 978-3-7315-0271-5

Band 15 Philipp Jung
Nonlinear Effects in Superconducting Quantum Interference Meta-Atoms.
ISBN 978-3-7315-0294-4

Band 16 Sebastian Probst
Hybrid quantum system based on rare earth doped crystals.
ISBN 978-3-7315-0345-3

Band 17 Wolfram Kittler
Magnetische Anisotropie und Quantenphasenübergang in $CeTi_{1-x}V_xGe_3$.
ISBN 978-3-7315-0363-7

Band 18 Moritz Peter
Towards magnetic resonance in scanning tunneling microscopy using heterodyne detection.
ISBN 978-3-7315-0410-8

Band 19 Junji Tobias Märkl
Investigation of Magnetic Adatoms with Scanning Tunneling Techniques.
ISBN 978-3-7315-0435-1

Band 20 Jochen Braumüller
Quantum simulation experiments with superconducting circuits.
ISBN 978-3-7315-0780-2

Band 21 Jinjie Chen
Local Investigation of Single Magnetic Molecules with Scanning Tunneling Microscopy.
ISBN 978-3-7315-0819-9

Band 22 Arnold Seiler
Einfluss der Leitungselektronen auf die Dynamik atomarer Tunnelsysteme in ungeordneten Festkörpern: Relaxationsprozesse in metallischen Gläsern und ungeordneten dünnen Aluminiumoxid-Schichten.
ISBN 978-3-7315-0870-0

Band 23 Jasmin Maria Jandke
Elastic and Inelastic Scanning Tunneling Spectroscopy on Iron-Based Superconductors.
ISBN 978-3-7315-0747-5

Band 24 Kevin Edelmann
Electroluminescence from Plasmonic Excitations in a Scanning Tunnelling Microscope.
ISBN 978-3-7315-0923-3

Band 25 Lei Zhang
Sub-Kelvin scanning tunneling microscopy on magnetic molecules.
ISBN 978-3-86644-950-3

Band 26 Lukas Grünhaupt
Granular aluminium superinductors.
ISBN 978-3-7315-0956-1

Band 27 Alexander Bilmes
Resolving locations of defects in superconducting transmon qubits.
ISBN 978-3-7315-0967-7

Band 28 Marco Pfirrmann
Adding nonlinearity to an electromagnetic-magnonic quantum hybrid device.
ISBN 978-3-7315-1003-1

Band 29 Andre Schneider
Quantum Sensing Experiments with Superconducting Qubits.
ISBN 978-3-7315-1032-1

Band 30 Vibhuti Narayan Rai
Light Emission from Single Self-decoupled Molecules in a Scanning Tunnelling Microscope.
ISBN 978-3-7315-1201-1

Band 31 Julian Antoni Skolaut
Molecular Motor Based on Single Chiral Tripodal Molecules Studied with STM.
ISBN 978-3-7315-1247-9

Band 32 Thomas Gozlinski
Local Probing of a Superconductor's Quasiparticles and Bosonic Excitations with a Scanning Tunnelling Microscope.
ISBN 978-3-7315-1279-0

Band 33 Matthias Micha Wildermuth
Quantum Tunneling of Josephson Vortices in High-Impedance Long Junctions.
ISBN 978-3-7315-1303-2

Band 34 Martin Spiecker
Two-level system hyperpolarization with a quantum Szilard engine.
ISBN 978-3-7315-1415-2

Die Bände sind unter www.ksp.kit.edu als PDF frei verfügbar oder als Druckausgabe bestellbar.